D0548403

Sun, Sand and Soul

Also by Lionel Blue
My Affair with Christianity

Also by Jonathan Magonet
The Explorer's Guide to Judaism

LIONEL BLUE
& JONATHAN MAGONET

Hodder & Stoughton
LONDON SYDNEY AUCKLAND

First published in Great Britain 1999

10 9 8 7 6 5 4 3 2 1

British Library Cataloguing in Publication Data
A record for this book is available from the British Library

ISBN 0 340 66926 8

Typeset by Avon Dataset Ltd, Bidford-on-Avon, Warks

Printed and bound in Great Britain by
Clays Ltd, St Ives plc

Hodder & Stoughton Ltd
A Division of Hodder Headline PLC
338 Euston Road
London NW1 3BH

To Doro, Gavriel and Avigail.

Jonathan

To my good friends who shared their holiday experiences with me:
Daphne, Derek and Gabrielle, Inge and Cyril,
Gordon, Guy and Leslie,
my Dutch friends Henk and Theo
– and not least Jim.

Lionel

Contents

Why holidays? Why this book?

For many years I've given short religious talks in the middle of a hot news programme of the BBC. These talks, sermonettes, Pauses for Thought, 'godslots', Thoughts for the Day, were not meant to deal with the complexities of higher mysticism. They couldn't – there just wasn't time. In any case the listeners, my 'congregation', were in bed or having a bath, shaving or swearing blue murder in a traffic jam and not disposed to hear higher mysticism or complex theology. (They just turned the knob or switched off.) Instead such talks had to be honest, relevant, personal, practical and spiritually helpful, fitting in to the rush of the social, political and financial life going on around them. They were at the cutting-edge where the life of this natural world met the supernatural world beyond.

In prayer I came to recognise that my first duty was to help people get out of bed in the middle of winter or in a recession and give them enough spiritual strength so that they didn't dive back under their duvet again. One thought was enough for one day and I fixed it with a joke because using religion to depress people is taking unfair advantage, and cheap, hitting men and women when they are down.

Some ecclesiastics made light of these talks, for they appeared to be light, but I took them seriously. The rabbis had constructed a holy way of life out of such small domesticities; Vivekananda did not despise wisdom from a dancing girl; and Jesus revealed the kingdom of heaven in a lost penny, a pot of perfume or in a lamp hidden under a bushel.

The teaching was two-way. Religion didn't know it all. It had to learn how to come out of its ghetto and deal not only with Truth, Love and Retreats but honesty, sex and package holidays. Ordinary people had been transposing the high-style stuff for years into the practicalities of supermarkets, the cinema, their pensions and savings. From their letters and chance meetings I learnt a lot. They provided my education in life studies.

In a synagogue, chapel or church you sometimes assume (I certainly did) that God lived in the ark or altar and a bit of Him (or Her or It) dribbled down to His chosen ministers, who then dribbled a bit down to the general public sitting in their pews. I now relearnt that God is everywhere, in any of us who let Him in. He is in the chap or girl you dance with at the hotel barbecue, in the holiday brochure as well as in the Bible, in you despondent in the departure lounge and in you in bed doing whatever you're doing. I was brought back to basics with a bang.

The aim of this book therefore is basic. It is to help you have a happy or happier holiday by sorting out your muddled metaphysics and not confusing comfort with happiness or too much cheap booze with self-acceptance and contentment. So Jonathan and I have tried not to exclude anything – your soul, condoms, Culture and Kulcha, Karioke and credit cards – even the nightmare when you've lost them. We have tried to analyse loneliness in five-star hotels, and loneliness in a crowd, vocations and vacations. We have tried not just to bewail them or comment on them but to give some practical tips about how to improve your lot and deal with them. We have tried to deal with you as you are.

We have put our names to our own life experience and holiday experiences. We have mused over them for many years, on the radio, in lectures, in books and journals – the BBC, *Country Life*, the *Tablet* etc.

We became hooked on holidays because they are so important in life and becoming more important each year as millions of us in Europe take part in the great north-to-south migrations. In grey times, many of us live on the thought of them. In holidays, romance and reality meet and 'somewhere over the rainbow' seems attainable. Why shouldn't it?

When my father was dying, he and I used to spend a happy visiting session poring over holiday brochures and planning holidays neither of us would ever make. This made him more tranquil than tablets and brought him sweetly to another world. (He had never had the money to go abroad.)

Like many of you, I've never had a lot of money either, though more than my father, so that wisdom has been learnt packaged off to Benidorm and Torremolinos, to Malta, Majorca, or Amsterdam, and in the cheap months too. But wisdom is no respecter of prices or places. So it's applicable in the Negresco in Nice, backpacking in Queensland, or paddling in Skegness. Some of it is applicable, too, on silent retreats, or pilgrimages with all mod con. But either way or any way, have a

Happy Holiday!

Life is tough and you deserve it!

LB

Nostalgia

I'd like to be a tripper
and eat a juicy kipper
on the Isle of Man.

– A memory of George Formby strumming on his ukelele.

LB

On the way

The brochure

In the autumn I gather another crop of brochures at the travel agents. They make a delightful winter read. I spend many mini-holidays lying in bed poring over them. I recommend them before going to sleep. The anticipation can be more satisfying than the reality. Each item has an in-built happy ending, just like the recipes in chatty cook books. Lissom youngsters or well-turned-out oldies reflect back to us the image we like to have of ourselves. The skies are blue, the palm trees green, and the buffet bars groan, supervised by helpful, smiling chefs. In the cold chill of London, can they be true? Is heaven a place on a package?

In the terms of the brochure it is, and not that expensive either if you avoid school holidays and hunt for the cheap packages. If they say you will get HB, SV, Balc., etc., you will be unlucky if you don't. Some brochures now even indicate noise levels and drawbacks, so you don't even have to complete the text with what it doesn't say. Generally speaking you're onto a good thing, especially if you've avoided the highest of high season! You have, of course, to keep your common sense. Eternal youth you won't get, a sun tan you will; comfort – probably, contentment – possibly, because that depends on you and you are not so reliable as the brochure.

So why has the promise of previous holidays not always materialised? Why have you arrived back home feeling unfulfilled, in tears even? You may want to blame the brochure and occasionally you may be right. Perhaps the sky was not as blue as the picture painted, or the indoor pool little more than a bath. I was once given the bridal suite on the fourteenth floor, where you flushed the veined marble loo with a plastic bucket and the lift was too tired to heave itself up to my level. Perhaps unscheduled construction works next door leave you covered in grey concrete dust, deprived of sleep and shaky.

If that happens make yourself a nuisance and don't be pacified by sweet words. A friend of mine was told he must sleep with two other males because the rep's poor old mother needed his room. Would he turn out her poor old mother she asked? Yes he would, he answered brightly, and what's more would sleep in the foyer until he was suitably fixed up as per the brochure. The mother dematerialised because she was never heard of again. Don't get too saintly. Being a doormat doesn't help you spiritually, or anybody else.

So where does everything go wrong? It is because you and the brochure both connive at the same metaphysical confusion. How can you connive at the same metaphysical confusion if you don't even know what metaphysical means? It means in this case knowing that *comfort* and *happiness* are not the same. The former concerns objects outside you and the latter a state of mind inside you: often they coincide but sometimes they don't.

I cast my mind back to other periods of my life, to

other holidays. I've stayed in high-class hotels (by invitation) and in basic youth hostels. I've been very happy in hostels and sobbed my heart out in marbled hotels. I would far prefer, of course, to be unhappy in the marbled halls than a hostel, but I've learnt through experience that comfort and happiness are not the same.

As a spiritual exercise, remember real happiness has to satisfy all of you, not just one bit of you. And you really are a rather complex creation. You have a mind, a body and a soul. There's an awful comeback if you only concentrate on one bit of you and leave the others starved.

Here's one form the exercise can take. It's one I've used a lot over the years though I don't know where I got it from. Shut your eyes (not obligatory) and invite your body, mind and soul to a tea party. Get them sitting round some imaginary table. Then start the conversation and let each one have his or her say about what they each want from a holiday. Don't let two of them gang up against the third and suppress it. The whole of you needs that holiday not just a bit of you. You can organise this inner conference in bed or on a park bench or in the departure lounge, though that's leaving it rather late. It might be a very interesting conversation indeed.

Let the different parts of you consider comfort, ecstasy, contentment, joy, happiness, inner quiet, body relief, even bliss – when you feel hoisted to another dimension. Which of these can you expect from your friendly brochure? And if your brochure can't deliver, because it's concerned with hotels not heaven, how and where do you find what you want?

These different parts of you have got to sort out some basic choices. Do you want a vacation or a vocation? Are you intending fun, slap and tickle, romance, 'in love' or real love?

The pictured people in your brochure are too busy playing handball by that blue sea to bother themselves with such probing. But you are more complex and contradictory than any model in a glossy brochure. You probably want all this and heaven too. Don't we all. But can we get both? You'll have to find your own answer to that one. This rabbi doesn't know. He'd still like to have his cake and eat it.

A prayer

> May my holiday be the one I've always dreamt about, and when I'm there may it be as wonderful as when I still dreamt about it back home! (A prayer based on Peter Abelard's definition of heaven in his 'O quanta qualia')

Courage when complaining

> Thou shalt not nurse hatred against thy neighbour; thou shalt reprove him frankly, then thou shalt have no share in his guilt. (Leviticus 19:17)

In other words, don't boil inside over an injustice. Have it out! You'll feel much better.

LB

Seven deadly sins on holiday – a quick checklist

A new brochure has come out which is really plain speaking and doesn't say 'refreshing' when it means 'rowdy', or 'lively' when it means 'noisy and you won't get to sleep until the early hours'. I study it carefully like the Bible, interpreting not only what it says but what it doesn't say, which is even more important. Soon I shall be one of the packaged millions surging south, dreaming of sun and something else, on an undiscovered, exclusive Costa on the Med. – God help us! But brochures, however plain-speaking, aren't the cure for any disappointments, because we disappoint ourselves.

I recommend an old discipline. After you've collected your brochure sit on a park bench or in a chapel and ask yourself if anything inside you spoils your happiness. If you're not sure how to begin, run through a checklist of the seven deadly sins to find out if they are the culprits – an old-fashioned but an effective test. This is a good spiritual exercise.

Start with VANITY, the most innocent yet physically the most uncomfortable. Don't breathe in buying new jeans – the ones in daring pastel shades! On holidays you'll eat and drink more than at home and you'll end up like poor TV darts players whose pinnies can't hide their beer bellies. Also, who knows what will go pop at the hotel barbecue – a zip, the all-important top button? Choose something comfortable instead. People like comfortable partners. But if your pants do go pop, remain philosophical. Even in an age of women's lib, many need to be needed and will enjoy sewing you together behind a bush. So will some prospective gay partners. I speak as a man but the same applies to a woman and you can transpose the garments yourself.

GLUTTONY means not piling everything on your plate at the buffet, cementing the mess with mayonnaise. It's such a giveaway! You're either so old you remember the last war and consume anything in sight or you're insecure and stuff yourself because the opportunity may never come again. It's also self-defeating. I saw a man with a tottering mountain of food on his plate and then he came to the last item, the special paella of the week. His tragic features showed he just didn't dare add it on. The other diners stayed their forks and speculated on the human tragedy. He left the buffet without his paella a chastened man!

Mental LUST doesn't seem so serious today. Few hide what they feel as boobs and bottoms parade by. But the difference between what you feel and what you do is civilisation. Instead of complex theologies of sex, here are some simple practical summaries for your own happiness. Don't do in Benidorm what you'll regret in Basingstoke! Don't do the night before what you'll regret the morning after! Keep appointments! Avoid lies – even white ones!

PRIDE used to concern what you were, your class or

family. But now it's what you have that counts. Now, some fantasy of high life is part of any good holiday and most of us want a marble foyer or brandies at the bar to back it up. But don't let such frou frou seduce you into a hotel with two stars too many. You won't be able to wander nonchalantly into the à la carte with your posh new friends but have to return to the old set dinner instead because it's inclusive. The strain of putting on the style can be jolly hard work.

ENVY means totting up everybody else's package, to work out who got the best deal. But this becomes fiendishly difficult. In one you get free tea and biscuits and in another a lower supplement for SV (sea view). What's worth what? You grind your teeth in despair. And you won't be able to enjoy what you've got because you'll be so mad about what they've got and you haven't.

SLOTH means getting too heavy mentally and psychically as well as physically, lying in bed and letting everybody else do the work, like planning trips and buying tickets. Others will have to provide the sparkle, the entertainment, and they'll resent it. In the end they'll dump you. Why allow yourself to get so heavy inside?

AVARICE is a fancy way of being greedy, like an addiction. Some are addicted to food, some to sex, some to gambling, some to their neuroses, some to cigarettes, some to booze. It's a compulsion and most people need help to get out of it. There's Alcoholics Anonymous and Over-Eaters Anonymous and Gamblers Anonymous. There's bound to be an Anonymous to fit you. Join it! Perhaps there's a Holidays Anonymous. You might prefer to work at home. Holidays can be the hardest work of all.

There's another sin known as ACIDIE which is a kind of boredom, meaninglessness, listlessness. It often strikes at noon or in the afternoon. It's more common in affluent societies like ours than in the poorer ones of the past.

But of course for a happy holiday be prudent as well as virtuous. Like the family who booked into that gastronomic hostelry on the Riviera I've called 'Chez Rabbin Bleu'. 'How's the food?' asked a friend. 'Well after the calf died,' mused the guest, 'we had superb veal fricassee. And when the hen died, Poulet Henri Quatre. But yesterday Rabbin Bleu died and we went self-catering quick.'

Putting on the style

'And when you were in Rome, did you see the Pope himself?' the Catholic woman asked.

'See him! Why he especially invited us for tea,' replied her Jewish neighbour.

'And how did you find him?' breathed the awestruck woman.

'Him I liked, her I didn't,' replied her neighbour nonchalantly.

LB

Special offers

I'm a sucker for special offers. Somehow, though I know there's a catch (there's no such thing as a free lunch), I still fill in the forms, make the telephone calls, stick little bits of plastic onto bits of paper – and always send my guaranteed chance to win, exclusive to the two million people who live in North London, in the special personalised envelope within the two-week deadline for that extra bonus prize.

Despite years of perseverance I've only ever won those little special free gift booklets offering helpful gardening tips or handy first-aid hints around the house, or else some trinket on a chain. Worst of all, I'm still tempted by those invitations to 'not-really-a-timeshare' presentations. I'm sure there are many legitimate operations that give value for money. What I resent is the pressure some of them put you under. Like the following examples taken from different experiences over many years in different parts of the world.

The come-on can be flattering. You are one of a few special people in your area selected by computer who have been invited to a not-to-be-missed presentation. Moreover, just by turning up, with your partner, of course, you are guaranteed a free holiday with our compliments.

The details may change but the pattern is depressingly similar. The nice receptionist introduces you to the nice agent. I don't know why but they often start by explaining that they are actually new at the job, don't really approve of all the methods, but are personally very impressed by the quality of this particular product. Looking nervously over their shoulder they may even show you something from the brochure you aren't supposed to see, hinting at special prices or deals! It is utterly transparent, but I get suckered in every time. ('Don't worry,' I want to whisper, 'your secret is safe with me.')

For some reason you have to listen to the explanations for at least an hour before getting to see the video, which reminds me of the filler travelogues they used to show in the cinemas before they invented television. 'As the sun sinks slowly in the west we bid farewell to beautiful Zanzibar . . .'

Then comes coffee or tea and the terribly convincing chat from the senior person who floats from enthused couple to enthused couple. Once they even had a board on the wall so that you could applaud each pair who took the plunge for the golden special deal of the month or whatever. So far they have been terribly friendly. But now comes the urgent sales pitch. Somehow unless you can actually make the decision this very moment (with, of course, two weeks to change your mind and cancel the nice cheque you have to write out), you lose out on a very special, once-in-a-lifetime offer they are making. Sign up and you get the warm feeling of an extra round of applause from all the other happy campers when your name goes up on the board.

At some point the boss guy recognises that I am not going to bite – and suddenly the charm goes, an abrupt dismissal is in order. If I insist at least on collecting these marvellous tickets for a free trip, they turn out to be excellent value if my wife and I can exactly fit into the very limited dates that are available, and we must pay an extra supplement of so many hundreds of pounds that move us up from a single room in the basement to a double in sight of the sea.

It is probably just a failure of imagination on my part. They tell you, in another stage whisper, that the place they want you to buy, on the sixth floor of a condo miles from the sea in downtown Miami, is not really so nice – but that doesn't matter since you could swop it one year for a bungalow in Tahiti or a commuter hotel in Japan. I just can't cope with the puzzle of buying something so that you don't ever have to use it – and then you can leave it in your will to your children so that they don't have to use it either. When they add that you are not really buying space but buying time, I get dizzy. As the Psalmist advises, I'd rather use my time to get wisdom.

I suppose they have their job to do – to sell so many packages per session. And at the end of the day I can understand that their patience, and charm, wears thin. I want to pat them on the back, tell them better luck next time and hope they make their quota, but they've already disappeared to commiserate behind the office door, or kick the dog.

The message in all this is a reminder that what is a holiday for you is big business for others or their living, and it is easy to get caught up in promises or special offers that entice you to spend more than you ever intended before you even get out the front door. What you would never dream of accepting when in your 'business' mode may seem awfully tempting in your 'holiday' mode.

The exercise I suggest is very practical. Use your 'business' mode before you leave home to think out the practicalities of the holiday. Just how much do you really want to spend, including something extra for a special treat. If you are likely to be tempted by local agents selling you a dream vacation, immunise yourself by a trial run at one of those presentations so that you've got it out of your system before you leave home. It would be a sorry business to re-mortgage the house you do live in for the sake of a fantasy one you can never live in. Then leave all that business stuff behind and enjoy yourself – after all, you are supposed to be on holiday.

JM

Some thoughts

It is an old folk tale, but no less true. Simple Simon is on his way home from market with two eggs. While walking along he plans his entire future life. He will hatch the eggs and breed two chickens. Once the chickens have begun to produce eggs he will sell one of them together with the eggs and buy a lamb. He will raise the lamb and collect its wool for a few years, then sell it and buy a cow. When the cow has given a lot of milk he will sell it and buy a bull so that he can breed . . . As he is walking along and thinking all this out he trips over a stone – and drops the two eggs.

Abraham Ibn Ezra said: 'Plan for this world as if you were to live forever. Plan for the world to come as if you were to die tomorrow.'

Special offers

* * *

I hope this poem doesn't sound too cynical, but I wanted to get all my anger about the commercialisation of holidays out of my system. It seems to be inevitable, and much of what is on offer is actually pretty innocent and fun, and provides a living for a lot of people. So check off for yourself the things that really turn you off, and enjoy the many opportunities that are left.

Happy Hour

I'd missed the Tom Jones and Dolly Parton
lookalikes
though they did indeed lookalike
in last month's freebee resort magazine.
I successfully ran the gauntlet in the lobby,
avoiding:
'day trips round the island',
'see the coral reefs by submarine',
'swim with the dolphins',
'learn to scuba dive in three hours, only $75',
(paragliding looked tempting
till I thought about trying to land),
'time-share apartments at special rates
plus a one-time-only incentive offer
if you sign up at this very moment',
'discount dinners with complementary cocktails
and a reduced entry to the show
of native dancers,
glass-eater and steel-drum player
at the Yellow Banana Nightclub'.

I settled under the advertising umbrella

by the pool
(no diving, no watersports,
no open wounds allowed)
ignored the Bingo-caller
and piped zombie music,
sucked some ice and Rum
from my 'Miami Vice' cocktail,
'second one free at "Happy Hour" ',
and wrote a poem.

JM

A postscript

Buying or leasing on your own a little bit of heaven which you won't occupy most of the year is also a problem. Property is a problem when you are not there in person to look after it, which is why time-share is attractive, despite the hard sell. People I've met have told me the satisfaction it gives them year after year. Some 'weeks' are quite cheap and might suit you and your family. But do consult your own agent or lawyer about your wonderful week before you buy. Timeshare is no different from mysticism and spirituality. The spiritual message is that you need discernment and discrimination for all of them!

LB

Expectation

A principal cause of unhappiness on holiday is not the harm others do to us, but the harm we do to ourselves through excessive expectation.

To avoid the horrors on the front page of my weekend paper, I started at the back with the introductions column. The advertisers seemed a jolly lot, most, so they said, with GSOH (a good sense of humour) and healthy high-class hobbies. But the problem is not their words but the expectation those words arouse – which turns ordinary guys and girls in our minds into the only one who understands us, the solution to our life's problem, the only one we've ever longed for, our dream, a Spice girl and a Chippendale – and all this before we've met. A fantasy and a recipe for trouble.

Misguided expectation like this can also wreck your longed-for holiday. 'It will be heaven,' we say. And this may be true, because the gates of heaven are everywhere, at home or on holiday. We go through those gates whenever we do something generous for the sake of heaven. But remember the kingdom of heaven is not just the rest our body craves or temporary relief from the worry of work.

No work for a few days is fine. We have nothing to lose

but our chains, as Marx said. Those chains of work and habit hold us together, which is why I always take some work with me. Think of the fate of some who had too much leisure, who won the lottery, pools or sweepstake, or who had all the power they craved like Hitler, Mussolini or Stalin, and their dreadful ends.

Misplaced expectation can also reduce your religion to magic. All your prayers will not move the cosmos off course to suit your convenience. But they can move you to serve the cosmos – if that is the 'answer' you want.

In insecure times, too much expectation is invested in wonder holidays, wonder love affairs, wonder gurus, charismatic leaders, wonder-working rabbis, inspired preachers and living deities. Be careful where you invest your hopes and remember the greatest teacher is your own humble common sense and life experience.

A Jewish seeker hears of a rabbi who is like Moses, like Shakespeare, like an angel! He rushes to the synagogue to hear this marvel. The service was tedious beyond belief and the sermon incomprehensible. 'How dare you make such claims for your rabbi!' he cries indignantly. 'Look,' said the warden, 'the Bible says that Moses stammered, well so does he. Shakespeare didn't know a word of Hebrew – neither does he. And like an angel, our rabbi isn't even human.'

Our excessive expectation is also the greatest destroyer of our happiness. So for a contented holiday (and life), do your best with your holiday as it is and your holiday companions as they are, not as you would like them to be. (Unless, of course, both it and they are truly dreadful.) Know this world for what it is, not as perfection but a corridor, another stage on your soul's long journey home.

To protect myself spiritually I do this exercise regularly.

I recommend it to you.

Think of all the comforts on your holiday which you have got and over 98 per cent of your fellow human beings haven't. Enjoy them and thank God for them!

I sometimes remember my distant cousins who went up in the smoke of the concentration camps. They would have thought the comfort and affluence of my ordinary cheap package heaven. This dreadful comparison brings me back to common-sense gratitude.

LB

A cautionary tale

A man complained to his rabbi. 'I get no peace at home. I can't sleep. My wife talks. The children cry. The dog barks. I think I'll go mad. O rabbi, what shall I do?'

The rabbi stroked his beard. 'Buy a goat!' he said.

The man cannot believe his ears but he did as he was told.

A few days later he returned, eyes red with tiredness.

'O rabbi,' he said 'now it's even worse. When the children cry and the dog barks, the goat bleats. I can't stand it.'

'Buy a cow!' said the rabbi authoritatively. The man tottered away and did what he was told.

Two days later he returned distraught. 'I am going mad, rabbi,' he says. 'Now when the children cry, the dog barks, the goat bleats and the cow moos. I'm at the end of my tether.'

The man could hardly believe his ears when he heard the rabbi order him, 'Buy a rooster!'

The very next day he returned a broken man. 'Rabbi,

have mercy,' he pleaded. 'What with the children and the animals, I haven't had a wink of sleep. Be merciful, rabbi!'

'Very well,' said the rabbi, 'sell the goat, the cow and the rooster.'

A few days later the man rushed to the rabbi and thanked him. 'Without the goat, cow and rooster life is heaven. I've never slept so well in my life. Thank you, thank you!'

LB

Packing

My mother had no trouble with packing at all. She just threw in whatever came to hand, her hairpiece (usually worn askew), a bottle of scent bought at the chemist, a scoopful of undies (who's going to see them, darling, except in the dark), the trousers she used to wear as a warden in the war, a pair of shoes with Louis heels, and anything else she could rescue from the clothes basket. She then called out to me to sit on her case while she secured it with a mighty leather belt, once again from the war, the First World War, my father's war. Going to the airport she wore the most priceless article of all – her mink coat. (She belonged to a generation which was not yet aware of animal suffering.) No man had bought it for her, she declared. It was the fruit of her own hard work. Any hard-working girl could get one if she set her mind to it. As she got older she became smaller and her mink heavier till she also needed help to totter in it. It was like a mink tent. She had also acquired a white mink hat which made her look like Brer Rabbit. But because she wore everything with conviction men respected her and admired her and flocked around her to the annoyance of her better dressed contemporaries who only wore clothes which 'went with' each other, the loot of M&S. She patronised both Harrods and C&A which she called fondly 'Coats 'n Ats'.

She packed with the same simplicity as she made connie-onnie butties for the journey. I suppose she packed that way as a reaction to her mother. When I was going to be evacuated as a child, my grannie took over my regulation rucksack and filled it methodically with her necessities which were not those of the ministry that supervised us children. She was convinced I would starve or be damned with forbidden foods. So out came the vest and pants and in went roast chickens, packets of smoked salmon that my grandpa sliced for me, cream cheese, apple strudel and bottles of borsht. I tottered to the evacuation queue supported by my family, bitterly mortified lest I fall on my back.

For packing, always remember that perfection is not possible in this life, so however carefully you do it, there will always be something left out. Be philosophical!

You also have to be honest. You may think you've renounced sex for ever over here, but on holiday you may have very different thoughts. So consider condoms carefully (the female ones too), and goo and soft porn if that gets you going. And have two thoughts or three about fur coats like my ma's fur and hairy tweed hacking jackets. Yes, they might be the most expensive things in your wardrobe, but sun, sand and sangria will do them no good on the beach. All clothes you have to breathe in for will do you no good either. Buttons will burst, releasing more breasts and bottoms than you wish. Here are some tips.

There are some things that need checking, like your pills, spare pair of glasses, and for me a spirituality book. But even these you can get abroad. The pills often taste nicer in foreign parts, being coated with chocolate or flavoured with cinnamon.

You could buy your souvenirs in advance from a charity shop. Then you won't have to lug them back across a continent. Far cheaper, too, though mean! Examples – German beer mugs, Moroccan bird cages, peasant pottery, funny bottles filled with banana liqueur.

The three great problems with packing are to my mind in the mind – they stem from insecurity, anxiety or vanity. Articles you can get abroad – foreign parts are no longer that foreign but your mind and soul need consideration. You cannot help it: you will have to take yourself along with you whether you like it or not – your tensions, your self-pity, your depressions, your bad vibes, your envy. You cannot help packing them.

My mother with all her idiosyncracies was an unexpectedly shrewd woman. 'It's not what you carry in your case but what you carry in your mind that spoils a holiday,' she said after I laughed about her harum-scarum packing.

Now, some people try to suppress the muck inside them but it just blows up all over their friends. And some try sobbing into their spaghetti, but that makes it too salty. I discovered a shrine years ago where you could post your complaints to God in a letter box at the side of the altar. I often wonder who read my letters en route but it worked a treat. Complaining to God does work provided you're brave enough to tell Him just what you think of Him and this unfair world He created. But give Him time to form His reply in your mind – that's only fair too.

And talking of spaghetti I picked up a piece of wisdom in Italy that my mother would have liked. If you don't eat up all your spaghetti, an Italian momma says, 'By all the saints I'll kill you.' But a Yiddishe momma says, 'If you don't eat up all that spaghetti, by God in heaven I'll kill myself!' I think the Yiddishe momma wins on points, but that may be my Jewish prejudice.

A suggestion

Sit back before you finish packing and meditate over previous holidays. They will begin to reveal patterns. Concentrate on the tiresome situations, rehearse them in your mind and see how you can live with them more comfortably or even change them, though this is not easy in a formed adult. Can you break free of a tiresome pattern? If at first you don't succeed, try and try again. If you can identify any pattern, you are already half way to freeing yourself from it. That's when prayer helps.

LB

Plugging away

There are not many absolute essentials for your package holiday. A Swiss army knife can handle most cutting and cooking-related problems. I like to take my harmonicas in case I come across a guitarist who can play a twelve-bar blues in E – but that may not rate very high on your own list. Instant coffee is probably cheaper in your home supermarket and chocolate addicts will automatically know what to pack – but wrap it in plastic if you're going anywhere really hot.

Reading matter is another obvious item for some, and a portable light as well if you like to read in bed at night. (Mediterranean hotels assume that beds are for sleeping in or other activities where a dim light is more a help than a hindrance.) Electric plug adaptors, an alarm clock, let alone currency convertors and various thief protection aids to wear around your waist, stuff down your trouser-leg and possibly even elsewhere, are so commonplace you can hardly get out of the duty-free without buying them. And a portable kettle or water heater can be a great comfort.

But all of these are nothing compared to the one item guaranteed to be essential on the basic level of survival in a foreign environment – a portable, universal-sized sink plug.

Sun, Sand and Soul

I don't know why they never work (or even exist) in foreign hotels (or in British ones for that matter) – though maybe I stay in the wrong sort of hotels. Even if they are there, small, black, wrinkled and slightly worn at the side, you know from the way they just slip into the plug-hole that they'll never hold a drop of water in the basin, and wrapping them in an old sweet wrapper, or forcing a couple of elastic bands round them never works. That pathetic pile of damp soapy underwear in the sink in the morning mocks at all those detailed instructions about soaking for thirty minutes in hot water while Sudso goes to work.

There may be deliberation behind this seeming accident. The bath plug doesn't work because it's cheaper on the hot water supply if you're forced to take a shower. And there *is* a laundry service you're encouraged to use. But conspiracy theories apart – and I truly apologise if I've maligned that lovely little place you yourself stayed in last year somewhere on the Costa del Sol – if you don't take that spare plug with you, you will be sure to need it.

The trouble is that even with the most careful planning we are bound to forget to pack something. And, anyway, isn't that just like life? We never know what we will really need till we find out that it is missing. Life is simply not pre-packaged and not every hole we meet can be plugged. Things, and sadly people, have the habit of leaking away over the years. So is there a kind of spiritual plug we can take with us?

JM

Songs

I find consolation in songs. They may not tell you anything you did not already know, but the combination of words and music can cheer you up or give you a good cry. So a useful spiritual exercise is to collect a few songs that speak to you. Remember when things fall apart that 'life is just a bowl of cherries, don't take it serious, life's too mysterious'. Or, if you think no one's ever suffered a broken love affair like you're suffering now, 'you must remember this, a kiss is still a kiss, a sigh is still a sigh, the fundamental things apply as time goes by'. When you're 'drinking my friend to the end of a brief episode, make it one for your baby and one more for the road'. Or if 'there may be troubles ahead' you may just have to 'face the music and dance'. If your holiday is a washout because of the weather, try 'singing in the rain', and when you walk through that storm, 'keep your head up high', or even 'keep your sunny side up' and 'look for the silver lining'. The list is endless and however obvious the words or simple the tune, together they can sometimes take the edge off the pain – long enough perhaps to move on to the next step you have to make. Maybe that's another reason why in Tin Pan Alley the ones who sold the sheet music were called 'song pluggers'.

JM

A thought

Even those things that we think of as superfluous in the world, such as fleas, gnats and flies, even they are part of the creation of the world. God works

through everything, even a snake, a gnat or a frog.
(Genesis Rabbah)

How much more so a humble bathroom plug or a sentimental song.

LB

Departure lounge

'Your flight is delayed', says the prissy voice over the tannoy, 'by an hour, three hours, a day, a week, a month . . .' And all your prayers and wit cannot wash out one despairing word of it. 'We apologise', continues the prissy voice, 'for circumstances completely beyond our control.' Her voice has a dying fall and you'd like to strangle her. The airline is one slice of the sandwich and the airport's another and you're the squashed tomato in between. Still, such as it is, it's your holiday so how do you make the best of it?

Well, if you're a wise virgin (Matthew 25:1–13), you'll have come prepared with smoked salmon and tomato sandwiches made from nice fresh bread and a kitchen roll to clear up the mess. A bottle of sparkling wine and plastic champagne glasses from your local supermarket would add the final luxurious touch and then you needn't join the scrum around the bar. You could sip your sparkling (almost indistinguishable from the real thing if it's non-vintage), and feel superior. If you want to be superior spiritually as well as gastronomically you can offer some to your neighbours. It would be a good start to your holidays.

The wine will calm you down. You're bound to be jittery

when you cross from one reality to another. After you've calmed down, you can forgive the voice over the tannoy. Remember what your families had to put up with in times past. Say they were the snooty sort who came over with William the Conqueror. Have you examined Norman boats? They must have been sick over the side! My own grandparents came steerage from Russia, and I bet there was no luxurious place to powder their noses. This departure lounge would have been heaven to them, so don't complain.

What about cheering up your neighbour who is also sunk in gloom. He is going to Rome on an art tour, so tell him about the old Jewish woman who's being shown around the Sistine Chapel. 'It took Michaelangelo seven years to finish this ceiling,' says the guide pointing dramatically upwards. 'I've got a landlord like that too,' sighs the old lady. Even if your neighbour doesn't smile because he thinks you're insulting art, at least you've tried, and spiritually that's the important thing.

Refreshed you can now explore the shops and enjoy the fruits of mild consumerism. There's free squirts of priceless aftershave and perfume going, but don't take too many or you'll be politely chased away and that would be bad for your new jet-set image. And don't take out your annoyance on the sales girl. Remember she's not going away on holiday but has to smile all day at people like you.

Restore your status with a book. What about the Bible, but in the language of the country you're going to. Your Sunday school religion may be in tatters but you probably won't need a dictionary for the twenty-third psalm or the Lord's Prayer. Also, apart from the spiritual benefits (unquantifiable), it might be fun. Did you know that

'Behold O behemoth' in some French translations comes out as 'Voici l'hippopotame!' If the Bible's too much when you're feeling so fraught what about a good reliable Mills and Boone?

Spiritually I get two things from a departure lounge. The first is that life is like a departure lounge. You make yourself as comfortable as you can in it, you make acquaintances and friends. But then your number is called and off you go willy nilly. It is not your everlasting home. Where do you think that is – in another place, in another life, in another dimension?

A departure lounge is also very good for people-watching. God comes to people in different ways but for me and many others, He comes closest to us through other people. It works like this – seeing so many anxious fellow human beings, a wave of compassion rises up in me for them and suddenly I know we're not just bodies packaged to Benidorm or Torremolinos, but immortal souls on our way to paradise.

LB

Solemn departure thoughts

Without your consent you were born and without your consent you live and without your consent you die, and without your consent you will have to give an account and a reckoning before God. (*Sayings of the Fathers* and early Rabbis 4:29)

This whole world is just a narrow bridge, but the essential thing is not to be afraid. (Nachman of Bratslav)

Everything is given on pledge, and a net is spread for all living. The shop is open, and the shopkeeper gives credit, and the account is open and the hand writes, and whoever wishes to borrow may come and borrow. But the collectors go round every day, and exact payment from people with or without their consent, and their claims are justified and the judgment is a judgment of truth. *Yet everything is prepared for the feast!* (*Sayings of the Fathers* and early Rabbis 3:20)

Some not-so-solemn thoughts

Try my mother's traditional departure lounge banana sandwich. Unzip a banana, lay it flat on a slice of buttered white blotting paper bread, then put another buttered slice of blotting paper bread on that, then bang the lot down with your fist. Fast, budget and delicious!

Why read other people's books when you could use the waiting time writing your own? Such as 'She trembled as she bit into her bap. She felt the rabbi's hot eyes piercing into her. She moaned as she munched.'

There are a lot of little goodnesses you can do which only God sees – like not pushing, or passing on your newspaper, or giving someone a sandwich or even a bite of your own. But the best is listening because delays make everybody tense. Listen to other people's remarks, they can be so funny, like: 'The hotel meals last year were poison, and they served such small portions too!' You listen to them and feel pleased you're human.

LB

Airport chapel

Hidden away in every self-respecting capitalist airport there is a chapel. They are not used much and are difficult to locate. Soviet airports of course never had them.

Sometimes the signs point to a chapel whose underuse has resulted in a takeover. In Spain the chapel became a temporary rest room for the airport staff. That was what it said on the door, but Information was most reluctant to admit the evidence in front of our eyes and I was too tired to squabble.

At Heathrow I'm told the chapel is underground but I've never found it or seen signs to it. Stansted was down a narrow corridor, but very satisfactory, and Gatwick showed signs of frequent use. Charter flighters used chapels more naturally than jet-setters – they are not so caught up in glitz.

Amsterdam in the old days had a pleasant chapel between the diamond counter and the security police, which made access to it daunting. Since then the airport has grown so much and is in such a constant state of expansion I have given up trying to find it. God knows where it is. I say this as fact not as exclamation.

In one chapel I made an experiment. There was a shelf of all the world's scriptures and I turned every third one upside down. Six months later I checked the shelf. They were exactly as I had left them. Why had traditional scriptures ceased to relate to modern problems of which there are many at airports in the hurly-burly and anxiety of arrivals and departures?

I meditated on this in the chapel hush and decided that another newer scripture is needed to help the old ones speak – a scripture which does not deal with 'them then' but with 'me now'. Unless you have located the divine in your own life, it is no use looking for it in Moab or Canaan four thousand years back.

So, as your spiritual exercise, locate the chapel and then locate the Divine in you. And instead of giving way to gush, start composing your own scripture – the religious story of your own life.

You can start practically by tracking the spirit down on your holiday because you will have the time. If you have a rosary or prayer beads, use them to count and fix in your memory your own religious experiences. If you don't, they will evaporate and cease to be real to you because they are see-through and fragile. They need not be startling but very ordinary – like stepping out of your body at a crowded cocktail party when you are holding a piece of asparagus in one hand and a glass of Spanish brandy in the other, and then from somewhere near the ceiling looking down on yourself and all the others with a deep and rare compassion.

They can be startling too, but to you alone. Perhaps the Divine will become real to you on the beach or become present in a bar and you realise that you are falling in love with love. Perhaps you also realise your home is in another world that comprehends more than this one, bounded by three dimensions of space and one of time, and that this world is only a departure lounge. Your eternal home is elsewhere and it pulls you towards it. Perhaps a wave of kindness overwhelms you and you do kind things, compassionate things that startle you.

Hold on to such experiences because then you will save yourself from unnecessary anxiety and heartbreak. You realise then that you cannot demand of this world what it cannot give – like permanence, and love itself. You get a taste of them in this world, but the fullness only when you have reached its borders. You may not think so, but such knowledge will make you more content and you will not pray for impossible things.

I did have one emotional experience in a chapel. I entered and unusually a group of people were sitting in the front rows. After a silence they spontaneously began to croon 'Somewhere over the Rainbow', that song of Judy Garland's from *The Wizard of Oz*. I was so moved by the sincerity of their voices, I asked who they were. They were of all beliefs and none they said, mainly none. But they all had terminal illnesses and had returned from Lourdes. 'Did you experience a miracle there?' I asked. 'Of course not,' they answered and giggled. 'Miracles didn't happen to old reprobates like us.' They don't happen to me either but their singing, though not a miracle, was a wonder and enough to send me off on a holiday a little drunk on joy.

Some advice on prayer. Don't determine in advance what's going to happen to you in the chapel, if anything. Sometimes you only feel the effect of your silence far in the future because God works in you deeper than consciousness. You don't have to do anything. There's another being involved in your prayers and it takes two to tango!

Sun, Sand and Soul

Sometimes people do better in prayer in the departure lounge than in the chapel. That's because there's less expectation.

Don't make things happen. You're on holiday. This time God can come to you.

LB

Guts for garters at the check-in

In the queue at the airport check-in, a lady asks me if I'm that nice rabbi, which promptly makes me feel nicer, for we all like people who like us.

'What a nice woman!' I say and only after mutual compliments do I realise she's wriggled in front of me during these pleasantries and is now five people before me in the queue. She smiles sweetly at the angry people behind her but it cuts no ice. They've seen that trick before. But I force myself to smile back nicely, though I am seething too, because nice rabbis can't tell her they'd like to have her guts for garters.

Actually, that is what I should have done because if you don't express your anger but suppress it, it will only fester inside you, causing you migraine or depression. After all, there are lots of queues on any package holiday. This is the first, and I ruefully realise I've set a bad precedent. There will be queues when you're marshalled to your buses, when you're allotted your room at the hotel desk, when you're selected for first or second sitting at dinner. Now some of these queues might really make a difference to your holiday, unlike this one which doesn't matter at all because you will all get on the plane somehow. No sensible package company is likely to leave you stranded at an airport, fuming and ringing your solicitor.

It isn't as if my niceness is religious, because real religion isn't cowardice. The Bible says explicitly when you've got a complaint against your neighbour point it out firmly. Which is why so much anger is expressed, not suppressed, in the Bible. Some of the psalms get so angry they have to be truncated or just left out of the liturgy. The prophets get really mad, too, and Jesus causes mayhem among the money-changers.

If suppressed anger is so dangerous, how do you get it out of your system? This is important because you bring a lot of suppressed anger from back home to your holiday. It's the reason why you're having a holiday, to have time and space to free yourself from it.

So before your anger takes control and sours your holiday and your relationships with your new companions, how do you deal with it? First, examine it because it might not be what it seems. Ask yourself if you've ever tried the same sort of trick and cheated in queues, because we often only get angry with other people for something in them we dislike in ourselves. I've also learnt from years of therapy to trace my anger back from this particular holiday to my earliest childhood, long before holidays were ever packaged. Humiliating as it might seem, it might only be the result of bad potty habits.

But whatever causes your anger, there's a lot of energy locked up in it which could be used for better, more creative purposes. So don't waste it, use it. You can do something better with it than throwing crockery around

– after all this is a hotel not your home. You can launch into prayer with it. You can bash a lot of beans, cooked or baked, with it, turning them into a nourishing paté, which is very handy if you've gone self-catering or even half-board. You can paint with it (art therapy is very useful), write it out of your system or evaporate it with a story like this.

There was this composer on the edge of inspiration. Just as he had almost got it, his adoring mother knocked at the door. 'Darling,' she said, 'eat this chicken, it's good for your brains.' Suppressing his anger, he answers, 'Yes mom,' and steers her back to the door. Again inspiration is about to come when she re-enters. 'I forgot this soup,' she says sweetly. 'For God's sake get out!' he shouts in anger, and his mother retreats in tears. Again inspiration nearly strikes when there's a knock. 'I've hurt your feelings,' she sobs. He cannot contain himself, he pushes his crying mother out, throwing her soup through the window. Then, eureka, he rushes to his desk, frenziedly crooning his new masterpiece – 'My Yiddishe Momme,' he writes, 'I need her more than ever now.'

Don't blame yourself for your anger. You're not the only one with it. The people around you have got to get rid of their tensions too. Remember you've got to love yourself as well as them. If your anger is too great for all the suggestions I've made, then turn up the radio or TV and scream it to God. He can absorb it, and won't mind provided you don't do it at night and keep other people awake with it. Some of the prophets did just that and why should you think you're any better!

As a spiritual exercise look up all the angry bits in a Bible. If you haven't brought one, there's usually a Gideon one beside your bed. Read them out loud with feeling.

LB

Some useful yiddish curses

May he have a large business, and may he never have in stock what they ask for, and may they never ask for what he has in stock.

May you lose all your teeth except one so that you can still get toothache.

May you have a large family of poor relations when you win the lottery.

I was angry with my friend,
I told my wrath, my wrath did end.
I was angry with my foe,
I told it not, my wrath did grow. (William Blake)

The plane – unexpected lessons of travel

We were finally slotted into our seats in the package plane and I warily inspected the man and woman in my row. In our adequate but restricted Lebensraum, for the few hours of the flight, we were going to be more intimate than we wanted because I had a prostate problem and they were between me and the washroom. But the holiday was already elating me, transforming and purifying me. The bad vibes were fading and I was feeling more and more benign. I no longer had to deal with people or put them on committees or think up the tactful wise words they wanted of me as their minister. I need no longer flinch from them because they were problems but could at last look at them as people. I could listen to them. If I only turned slightly I could 'people-watch' them.

And 'people-watch' can easily become 'God-watch' I told myself excusingly – because I had thought of meditating or praying on the plane like some pious Anglo-Catholics I knew who always scampered away from life to liturgy if they got the chance. 'If I purified my motives and didn't do people-watch to make fun of my fellow travellers (though our fellow humans are indeed funny – like us), or to annoy them or to feel superior to them,' I told myself, 'my people-watch could become a godly exercise.' For many of us, the closest we ever come to God is spotting Him in other people. It helps us to go on to the next lesson and spot Him inside ourselves.

The man and woman in my row had already made contact while I was contorting myself into reasonable comfort and debating my phoney spiritual problem. They had done it quickly, but then they were singles and singles need someone more urgently than couples. Few of us feel complete alone. They were nice but preoccupied, especially the young man. He was already recounting with restricted gestures the heroic story of his own struggle against life's unfairness. As I've told the same tale many times myself, I was more interested in the girl beside him and how she managed to consume a burger and bun from her bag, while nodding in sympathy and skilfully throwing in 'ohs', 'ahs' and 'reallys' between bites. Listening and eating like that was an art, and she didn't drop a crumb. She must have raced to the airport without time for breakfast, but of that she never said a word.

I suddenly wondered how it felt being a woman and being expected to listen to men. Do they listen because they want to? Or is it their nature? Or the way we programme them as babies? And my mind roamed further to the parts never mentioned in sermons. What was it like having periods? And don't they ever want to reverse roles and gesticulate too and talk about themselves non-stop while a man smiled in sympathy and tut-tutted? And how did they feel when they saw all those conferences on the

TV conducted by macho males locking their horns like elks while the women had to clear up the mess? How did they endure being patronised? Did women hunger for men sexually as men hungered after them? Did they have the same fantasies about each other? What did women really think of men?

Not much according to my late mother and aunt. In the nursing home where my friend and I used to visit them in the last year of their lives, we noticed how much they talked about meeting their mother, great aunt and sister in heaven but seemed surprised when we pointed out that their father and husbands would await them too. They didn't mind but men just seemed unreal.

'I'll tell you a joke,' I said, and they smiled in anticipation. 'A little boy says to his mother at dinner, "Isn't it a shame ma that that poor dumb ox has to suffer so much so that I can enjoy this little fillet steak?" "Please don't talk about your poor dear father like that, darling!" reprimanded his ma.' 'Lionel was a bright little boy like that, Hetty,' said my aunt fondly to my mother who nodded. They just couldn't see anything funny in my joke, though both their husbands had been good providers who worked as hard as oxen to keep their modest little households out of the hands of bailiffs in times of recession and slump.

I suddenly realised that though I've been concerned with marriage guidance, divorce and family matters all my working life, I've never really known what it's like to think as a woman in a woman's body. And the opposite is probably true too. Do women understand the expectations of power and success most males have to carry? They can't hide their tiredness with false orgasms – their lack of erection gives them away. Perhaps we all ought to go to evening classes and learn about each other, and I don't mean just sex but gender.

The text 'male and female created He them' (Genesis 1:27) came into my mind. On this holiday I would try to understand more and judge less. This would help me understand myself because the text should really be read 'male and female created He us'.

My meditation was interrupted as the little hot lunch trays were distributed. The man I was pleased to note offered the girl a little bottle of wine to accompany it which she tactfully refused. But after he had drunk his wine when he was beginning to mellow and untense, he began to ask the girl beside him at which hotel she was staying and what she did back home and whether she had the same problems in her office as he did in his, which was magnanimous of him. 'O my work isn't so interesting,' she said casually and tactfully, and he must have approved because he then invited her to his hotel barbecue. She had won the set and I noticed a speculative look in her eye as she gazed at him in between more 'ohs', 'ahs' and 'reallys'.

Some hard advice

When we are away from home, we see the obvious but with fresh eyes. At home we are too defended by domesticity to see – really see – what is in front of us. That is a good reason for travelling.

It is also a reason why we do not get on with companions we liked well enough at home. We had to cross a continent with them to see them as they really were and ourselves as we really are. This unfortunately also applies to spouses and lovers. With all that time together, we have to come to terms with each other's otherness.

LB

Passport

I still feel the chill of that first visit to East Berlin in the days when the Wall still stood. The guide on our coach on the Western side was smart, witty and attractive, as if to heighten the contrast with her drab and dumpy counterpart in the East. Her parting, well-practised, words offered little comfort.

'Don't worry, 90 per cent always come back!'

Of course, as Western tourists, there was nothing to worry about, and yet . . .

There was that strange moment when my passport disappeared through a slot in the grey wall and a shiver literally ran through me. Would it come back? Would I be trapped, identity-less, in this grey limbo of concrete and barbed wire between two worlds?

The moment passed of course. But it was just the slightest taste of what it must be like to live on the wrong side of that wall, imprisoned within a police state, identity lost or reduced to a number on someone's list.

So we spent the day in those drab streets, and I remember the sense of relief when that precious passport, duly stamped, was returned on departure. The magic of that stately assumption of authority and power had worked yet again:

> Her Britannic Majesty's Secretary of State Requests and requires in the Name of Her Majesty all those whom it may concern to allow the bearer to pass freely without let or hindrance and to afford the bearer such assistance and protection as may be necessary.

Soon such experiences will be remote in the 'New Europe'. We will have to travel further afield to experience that cold discomfort and nagging uncertainty. And the blank gazes of passers-by who sum up your Western clothes and calculate the risk of talking to you. They have no passport and no way out. Such a simple document. So easy to take for granted. Yet for some a matter of life and death.

Sadly you don't have to go abroad to find bully-boy tactics with defenceless people trying to enter a country. Some years ago I had to meet a friend arriving from Africa at the airport. She was facing a chaotic situation and managed to leave home without the proper visa. She also made a stop-over in Germany which was to add to her problems.

The officer at Passport Control would not let her enter the country, even when I offered to vouch for her. So I tried to explain how difficult it had been to communicate with her to make sure she had all her papers. 'It takes six weeks for mail to reach her,' I pointed out. He pounced on this like a demented Sherlock Holmes. 'She said it took four weeks!' Wow! He'd caught us in an untruth, now he could unravel the rest of our lies.

I blew my top and pulling as much rabbinic rank as I could muster demanded to see his superior officer. Eventually he appeared, even more authoritarian in his way, but smoother. His approach was to treat me as a rather silly little boy who did not really understand the significance of his task and my patriotic duty. He was the first line of defence against the miscreants, cheats and villains who were trying to flood our green and pleasant land. Nevertheless, out of the kindness of his heart he was prepared to grant a temporary visa for a week if I guaranteed she would turn up to be put on the plane out. I swallowed my anger, mustered as much humility as I could and accepted gracefully his generous offer. (I don't like moralistic sermons from fellow clergy, let alone customs officials.)

In the end he got his revenge. When I returned her to the airport a week later on time, instead of putting her on a plane directly back to Africa as we had arranged, they sent her to Germany – after all that was her last port of call before she arrived! So it needed a call to a pastor friend in Düsseldorf to look out for her and find her airfare home which I had to replace. Now all this happened years ago, and immigration officers and the rules they implement have become more sensitive and more caring. I am glad because my own grandparents came here as refugees from Russian pogroms. They could as easily have been sent back to their deaths if one of their papers was wrongly stamped or missing. This is the age of the refugee.

This subject doesn't easily fit in to a happy holiday book. But when you take out your passport, or pass without fuss through one of those little cubicles where uniformed people give you a bored or friendly nod, think of all those who queue endless days and nights in embassies around the world hoping to get that essential piece of paper that may make the difference between life and death.

And when the tabloids scream about the menace of asylum-seekers, get angry and write the paper a letter. The whole of our Jewish, Christian and Muslim traditions go back to a few refugees like Abraham and Moses, and a bunch of slaves who managed to escape from Egypt. As every traveller knows in his or her heart, 'We are all of us strangers almost everywhere in the world.'

JM

Bible thoughts

> When a stranger settles with you in your land, you shall not treat him harshly. Like one of your own homeborn shall he be to you, this stranger who settles with you, and you shall love him as yourself for you were strangers in the land of Egypt. I am the Eternal your God. (Leviticus 19:33–4)

> Do not hand over a slave to his master if he has sought asylum with you from his master. He shall live with you, amongst you, in a place of his own choosing in one of your city gates for his own good. You shall not treat him badly. (Deuteronomy 23:16–17)

An exercise

Imagine yourself in the same airports you pass through, not as a holidaymaker but as a refugee!

Queues

On some of the exits of the autobahn in Germany where the traffic is reduced to a crawl and may be held up a long time, even backing up down the autobahn itself, a sign reads: 'You are not *in* a queue, you *are* the queue.'

This is true, though preachy. It must also have a very negative effect on the poor drivers sitting there for hours on end contemplating the exhaust fumes of the car ahead as their own engine overheats. Still, the point is well taken. I am as much 'the queue' as anyone else. If none of us was there . . .

But since queues cannot be wished away, they have to be faced. Obviously our patience and tolerance depend on a lot of factors, such as how much time we have to spare or kill or waste. I say this as someone not yet converted to a portable telephone which has presumably abolished the queue as a problem, and converted it into a business opportunity.

My favourite queue, through the rosy lenses of the 'retrospectoscope', was in Moscow airport. I had to change there to a local flight, having arrived from Heathrow. The fun began when I assumed I should head for the transit lounge, but the nice lady in uniform guarding the entrance assured me I had to follow everyone else downstairs and pass through passport control. It seemed reasonable. I had an hour and a half before the next flight to Kiev. What could go wrong?

I found out at the bottom of the stairs. There were half a dozen booths at the far end of the long hall, with uniformed figures sitting in them. And theoretically six orderly queues leading to them. Except that the room had clearly accumulated a dozen flights already, and new groups arrived every ten minutes with the same happy expectation of being quickly processed. My queue, three wide, was subject to unexplained changes as others joined or departed on the other side of the pillar about five rows ahead of me. I got quite used to that pillar. I could measure my progress towards it, a foot every fifteen minutes, though it wasn't clear that reaching it made any difference to actually arriving at one of the booths.

You know the way it works. You look at your watch. You have ninety minutes. At one passenger every two minutes you should be through in thirty minutes maximum. Five minutes later you refine the calculation. At four minutes per customer it will take sixty minutes. Ten minutes later, when your own queue has not moved, but the one to the right of you seems to be making progress, you start to revise upwards. You now understand that Einstein must have developed the laws of relativity waiting in a bus queue in Berlin. Staying in my queue might now take the full ninety minutes – which was already down to sixty anyway. But will changing queues help, or will I only lose my place here? To frustration and indecision we can now add incipient panic.

At times like this I wish I had the unabashed shamelessness of my father. He would join a queue for a new movie in Leicester Square with my mother. As soon as the doors were open he would march up to the front to the usher, introduce himself as a doctor, explain that his wife was recovering from a very serious illness, and would it be possible to let him in first? I don't know how often he tried that stunt, but I never heard of it failing. He would then herd my mother, red-faced in embarrassment, hardly looking like the wan invalid he had described, up to the front of the queue and into the cinema, with a grateful smile and usually a half-a-crown tip. It was disgraceful behaviour, and I wish I had a quarter of his cheek – or do I?

Back in my own queue the suspicion was dawning that I would miss my flight and might be in the queue for ever – or worse still be trapped in Moscow Airport in some limbo on the other side of those nightmarish booths. It was time for 'plan B': find 'someone in authority'! One booth at the far side of the hall was unattended except when a group of obvious locals passed quickly through it. A uniformed lady was standing nearby. She did not seem particularly concerned about my predicament but that may be partly explained by the fact that she seemed to speak no English. Consulting two of her colleagues did not seem to help, so I trusted my intuition and headed back up to the arrival lounge. It was now empty, without even the guard to the transit lounge. Trying to look like I belonged I marched through. Do you know that piece by Franz Kafka? A man walks past the guard, pauses and returns: 'Does your silence mean that it is permitted to pass?'

I couldn't help remembering the 'good old days' of the Soviet Union when any number of KGB guards would have arrested, strip-searched and, for all I know, imprisoned me for a lesser infringement of Soviet space. I felt like Alice through the looking-glass.

The transit lounge office had a small glass window to address the assorted nationalities trying to get help. Fortified by the knowledge that I was half an hour short of missing my flight I barged to the front. Daddy would have been proud. It took a few minutes for my situation to register but a helpful lady emerged and took me through all sorts of backdoors to get my passport stamped and ticket checked. She even commandeered a bus to take me to the other end of the airport where the internal flights departed. She was efficient, apologetic and a tribute to the newly emerging Russian entrepreneurial type. I wish I could remember her name to thank her publicly. She assured me that my luggage, which was temporarily mislaid, would soon be aboard. I never saw it for the rest of my stay – but that is another story.

The spiritual lesson to be drawn from queuing is the value of patience and resignation. If you have done all you reasonably can to change the situation and you simply have to wait, then yield gracefully to fate. Get to know someone nearby who is also stuck. Or else . . .

In my worst moments I imagine the waiting room to heaven and hell is a bit like that basement room at Moscow airport. But with all eternity before us there will be less urgency to get past those never-nearing booths. It is also the one queue that, sooner or later, we all have to join.

JM

Lost luggage

Luggage halls come in different shapes and sizes but they generate a lot of excitement waiting for your case to appear – or anger if there's a go-slow in the system.

I like the smaller Mediterranean airports where cases snake through the arrival hall and disappear through a hole in the wall only to re-emerge a few minutes later with new companions. You can peep outside and see the trailers stacked high, hear the curses of the handlers and wince as they dump your delicate bags onto the belt.

Big airports are more anonymous. Cases emerge out of holes in the ground and arrive with a slap onto the belt or a crack when they slide down the metal slope and bang into the containing wall. Heathrow provides one of the steepest gradients so they land with the loudest crack. The indignity of it all.

Somewhere, in Germany I think, it was all much more precision-orchestrated. Cases did a kind of dance between three moving bands, changing direction each time, stopping and waiting till the way was clear before moving onto the next stage. A miracle of photo-electric cells and neurotic engineering created this stately dance.

But how long do you wait till you accept the awful truth that your own luggage has disappeared? By the time the same assorted remaining bits of luggage have gone round on the carousel a dozen times – cases that you remember were there before your own flight was unloaded – it is time to admit defeat. Somewhere between check-in desk B and arrival hall A your luggage has gone astray.

It is like having your car stolen. Getting over the sense of unbelief is the hardest part. You parade up and down the street where you are sure you had left it parked, embarrassed at making a fool of yourself for some oversight – was it really here or did I go round the corner? But when it is absolutely clear that the car, or luggage, is gone for ever, then the frenzied search begins for help, explanations, mingled with anger at some strange power that has chosen just this moment, on just this holiday, to mess things up.

When it happened to me I bumped into another problem. Buried somewhere in the small print of my ticket, I was told was the information that some airlines only had a limited liability for luggage that went astray. For some bizarre reason they could only pay compensation according to the weight of the missing luggage, not the contents. So they referred me to my insurance company if I wanted compensation for anything above a couple of hundred pounds (twenty kilos allowance times x pounds and y pence). But my insurance company recommended that I went back to the airline, and anyway pointed out that it was only prepared to compensate for missing contents for which I had a receipt of purchase. This cut their liability and elevated my blood pressure. To have got it right I would have had to make sure that I only packed new, receipted items in my luggage, or old ones I could afford to lose – plus a few rocks to make sure my case weighed the full kilo allowance. (Of course, if my luggage had not been lost, it would have been a bit difficult explaining the rocks when I went through customs.) The moral is: always check the small print and consult your travel agent because the situation may have changed for the better in some or even all airlines.

What is sad about the whole business – and fortunately it does not happen too often – is the need to put the same kind of careful planning into your packing that you would do in your working hours – just when you wanted to get away from such chores. So alongside sorting out your insurance, your EC form for medical treatment abroad, your new mini-passport, your international driving licence, your brochures, emergency telephone numbers, tickets, five reduced-rate vouchers for the Casino Roman party, with a free glass of champagne, and the address of that lovely bar on the front that Jack and Lizzie visited last year, medication that you need, medication that you might need, medication that you don't need but always take with you anyway, travellers cheques, spare cash hidden in a wallet on a chain stuffed down your trousers and a secret emergency one in your sock, you have to make sure you have itemised, costed, kept duplicate receipts (originals only accepted by the insurers!) and notarised every single thing you are packing.

Once you have got over the shock and explained it several times to assorted agents, airline staff, travelling companions and the resident airport spiritual counsellor, it is time to take stock. And it can be quite a salutary, not to say spiritual exercise. What do you really need to survive the next couple of weeks? A few toiletries, a couple of changes of underwear and socks and soap powder, one extra shirt or tee-shirt, depending on where you are staying – and that is probably enough to be going on with. (I presume there are equivalent necessities for women but would not know where to begin to name them.)

Suddenly the several changes of clothing, electrical gizmos, reading matter, exotic bath salts, funny hat you bought two years ago in Ibiza and three pairs of swimming trunks, don't seem to matter so much. No one can expect you to dress up any more. In fact you are more than likely to be the recipient of all sorts of useful additions to your wardrobe from sympathetic friends. Suddenly you are free from the conventions of holiday overload.

In short, as long as you keep a few bare essentials in your hand luggage, it is possible to survive without all the paraphernalia, and the whole thing can be very liberating.

JM

A prayer and a thought

My own personal traveller's prayer: May my luggage return to me undamaged, unopened, at the proper time and place, and may I get to heaven more or less intact as well.

The great medieval philosopher Moses Maimonides pointed out that the things we really need are very few; the things we don't need are infinite, so our desire for them is infinite!

JM

Losing things!

My great, great aunt died and my grandfather took me when I was a child to the house of mourning. 'I know exactly how you feel,' he said benignly to the mourners. 'I hate losing things.' On the way home he gave me a long lecture about tact with the bereaved.

LB

Journeys

Away from here

I had travelled to Margate on August bank holidays and once to Calais on the Golden Eagle for a day trip. But these were my parents' holidays. My own began when the Second World War ended. I took my rucksack to school on the last day of term and as soon as the drawn-out school assembly ended, I made my way to the Old Kent Road and hitched to Rochester, taking the old well-trodden pilgrim route though I didn't know it. (Bowed down by the weight of my rucksack I must have looked like a medieval penitent.) And beyond Rochester was the mysterious continent so recently liberated, the Mediterranean of desire and the Paris of wit, sophistication and whatever (lots of whatever).

'Here' was the characterless suburb which I inhabited but wasn't home. It was the adolescent pimples and spots which cratered my face. 'There' was passion, fate, horror (the concentration camps had been gassing kids like me not so many months before), the hope of bedding another human being (preferably male but female would do). I was open to anything and up for grabs. Looking back, I was lucky to have returned healthy and disease-free and without a criminal record, but these were pre-pill days before dope, AIDS and drugs. I didn't realise my luck at the time but you have to acquire experience and wisdom before you see the obvious. Sometimes I joined up with a fellow hiker but I was too prickly for our relationship to endure.

I was becoming pricklier and pricklier with self-pity because I never found what I was looking for, only what I wasn't looking for and couldn't appreciate. Like the divine love of a worker priest in Abbeville, or the boundless, crazed compassion of a resistance lady (women were still ladies to me) who put me up in the turret of a ruined castle, or the cynical ex-German soldier on the run who shared his baguette and cigarette with me (he only had one) and his bitter life experience, or the poor Jewish woman teacher who gave me her lunch money for a meal. They gave me a self-knowledge and feelings which I would have preferred to disown but couldn't. I didn't want their depth or reality or self-sacrifice, but kitsch – continental film kitsch.

Lying in wait along dusty roads for compassionate car drivers I began to realise my outer journey was turning into an inner journey, though I didn't want it. Without willing it, my scrabblings over the map were propelling me through muddled despair to another dimension.

When I arrived back after nearly two months' wandering, my parents noticed something was not right and asked me tentatively if I'd had a 'nice holiday', because I was certainly stranger than when I had set out. 'Nice'? – I didn't know what they meant! I was even closer to a breakdown than before and in a few years would have it. Also I realised that I had to get 'away from here' to grow, and that for me, as for many others, an outer journey would always turn into the prelude to an inner one. I just couldn't be a pagan, however much I tried.

It was difficult explaining this to myself, let alone to my parents or other people – and made especially hard because at that time I was a materialist and not spiritual at all. I found some solace and companionship in books. I started thumbing through the *Pilgrim's Progress* I had encountered in RI lessons in childhood. I had grown towards it because it wasn't arid but alive. Later on I realised my affinity with stubborn, disordered, mystical Margery Kempe, and even later with the tramps of *Waiting for Godot.*

During my psycho-analysis I began to bring the different parts of me together and tried to live my own life not other people's. That is how I came to put the outer and the inner journeys together, discerning spiritual truths in package holidays and how you dance with your soul in a disco.

Only in my sixties did I learn how to have such a holiday. I hope you can come to such knowledge sooner.

LB

A prayer for first-time voyagers

Protect me, God, because my boat is so small and Your ocean so vast. (French prayer)

Bunyan – the greatest inner journey of them all

There is not much point in just covering mile after mile of space. I found this out when I went hitchhiking after the war. I became obsessed with distance – how many miles I could cover – ignoring the country and beauty I was passing through. Many post-war hitchhikers were like me, compulsive travellers. It was John Bunyan, the English tinker at the time of the Restoration, who made my wanderlust sensible by connecting my journeys over the map with my inner journey. I wasn't just travelling to the next youth hostel, I was also journeying to a country 'far beyond the stars'.

I first came across John Bunyan when I was evacuated and the only Jewish boy at a non-Jewish school. I was curious about Christianity and my parents finally gave in and I was allowed to attend school assembly with its prayers and plangent hymns and Religious Instruction where we didn't learn the Gospels, whose miracles I never believed in (I didn't believe in any, irrespective of race, religion or creed), but the *Pilgrim's Progress*. Because of the names of the characters I caught on straightaway that the progress was not just through Bedfordshire but through life. I had just started to identify the characters (Giant Despair and Mr Valiant for Truth) in real life and in me when I was taken back to London for the Blitz – my parents wanted me to be with them in case the invasion took place, which we expected any day.

I never thought much about Bunyan until one day during a difficult adolescence I took down *Pilgrim's Progress* from a shelf and the first words riveted me.

> As I walked through the wilderness of this world, I lighted on a certain place where was a den, and laid me down in that place to sleep; and as I slept I dreamed a dream. I dreamed and behold I saw a man clothed in rags standing in a certain place, with his face from his own house, a book in his hand, and a great burden on his back, and read therein; and as he read, he wept and trembled; and not being able longer to contain, he brake out with a lamentable cry, saying What shall I do? What shall I do?

I was crying the same question inside in my adolescent wilderness and after that my progress and the pilgrim's progress joined up. I packed my second-hand Victorian copy in my rucksack and read chapters from it as I lurked in hedges along the dusty roads of France, thumbing my way to the Med. Its pages were stained with red because I could only afford yards of baguette moistened with squashy tomatoes.

I finally got to Bedford because there used to be a farm there where young left-wing Jews trained for an agricultural collective life in Israel and I used to visit a friend. While she worked in the fields, I wandered round the

Baptist church examining the stained-glass windows with their scenes from the book. I stared at the one where the pilgrim unburdens himself of his rucksack and I wondered how I could unburden myself of my depressions, lies and repressed sex. You needed faith, a slippery commodity. I didn't know where to get it or how to hold it if I did get it.

I also stood beside the River Ouse, Bunyan's river of death, staring at the far side, Bunyan's celestial city, and wondered how and if I would ever make it there. I finally identified not with Christian and Mr Valiant for Truth, for whom 'trumpets blew on the other side', but with Miss Melancholy with her strange songs. My life was pretty knotted and strange at that time too. My meditation was ended abruptly as I nearly got bitten by a swan.

I also felt the weight of Bunyan's old anvil, the actual one he carried on his back through the hamlets of Bedford looking for work – he was a tinker, like some of my relatives in Russia. It was so weighty I couldn't lift it and my heart went out to that poor steadfast man, enduring year after year in gaol, about to be transported to the Indies, and worrying who would care for his blind daughter then.

Studying history I realised that he was also bound by his time, and this released me from ascribing perfection to anybody. He was a foe of the Quakers, my religious mentors, and believed in witchcraft when poor old women were tortured and murdered all over England. He also wrote a sermon on unbaptised babies in limbo, which I had no wish to read twice. Still, he taught me the great truth about journeys. We weren't just bodies packed in buses, we were also souls helping each other on the way to heaven and the celestial city.

I repeat this truth to myself even now as I go to work on London's underground, or on a grey day hurrying to the BBC, or leafing through brochures planning two weeks packaged in an earthly paradise.

LB

Something to sing

He that is down need fear no fall;
    He that is low no pride:
He that is humble ever shall
    Have God to be his guide.

I am content with what I have
    Little be it or much
And Lord contentment still I crave
    Because Thou savest such.

Fulness to such a burden is,
    That go on pilgrimage;
Here little, and hereafter bliss
    Is best from age to age. (John Bunyan)

As I travel through the bad and good
    Keep me travelling the way I should.
Where I see no place to go
    You'll be showing me the way I know.
(Sidney Carter)

Bunyan – the greatest inner journey of them all

As I wander after the elephant
　　People tell me where he's been
But I never can discover
　　If he's the elephant that I mean.
　　　　　　(Jonathan Magonet)

My soul needs a holiday too

I'm off to Berlin on business, then back to London, and then on to Cyprus for my longed-for holiday with barely a day between to pack fresh undies, if I'm lucky. And then old calamity strikes up again. That inner voice (I'm too cross to award it a capital letter), dormant during the last three weeks, whispers, 'You've just got time for a holy day with the Carmelites, if you're quick!'

Partly pleased that my inner voice and soul are still around, and partly sulky, I ring up the Carmelites to see if they've got room and will have me (perhaps they won't, which will be simpler all round). But they have and will, so I can't retreat from my retreat now (dammit, Paddington is not the easiest station to get to!). Actually I know they'll always have room and even make room for me, which is one of the nice things in my life.

Really it's all rather silly, because I can pray anywhere. There's lots of local places of worship – all brands, all flavours – which would love to have me. So why wander out into the country, which I've never liked. It's a lugubrious place, dripping trees and dead bird corpses in the copses. In any case my inner voice is all 'as-if' stuff.

Which is true, but I still get out my holiday rucksack with a sigh, which is answer enough. Now is not the time to give my voice notice to quit in my mind (I'm an OAP and may need it). 'So stop agonising Blue!' I admonish myself. 'The rucksack's packed, its non-kosher contents weeded out and the die is cast.' My partner also says he'll run me to Paddington which is nice of him, because he's off religion at the moment. (Having been an undertaker he's always had doubts.) And then my piety begins to perk me up. There'll be the friars who fulfil the role of the brothers I always wanted. And then, of course, there's that voice in the silence, and I wonder what He/She/It has got up Her/Its/His sleeve. Perhaps it will be nice like a new box of chocolates with lots of soft centres. I always expected religion to be tortured and agonised, but I've enjoyed it and have had to stop myself getting high on it. Why are my nights not dark enough? Have I plugged in to the wrong hole maybe? God knows!

It's nice to go to a place that's custom-built for the inner conversation, where there's more silence than liturgy. (I've run too many laps round too many liturgies, and they've all melted together like chocolates left in my trouser pocket.) The Carmelites also know the know-how of the inner life. Silent prayer makes me famished and I need stodge. So there are digestive biscuits distributed in sacred nooks. If you're going to dive into the Nothing and Nada like St John of the Cross, the black hole within you needs a good lining of comfort food.

It's quite confusing if you're a Jew fallen among Carmelites, even though a Jew with a lot of stretch. One friar said to me, 'I didn't realise how Christo-centric we were until you popped in'. Which is sensitive of him.

You do have to be pretty nippy about sizing up a prayer to see if you can recite it, or whether it requires so many – or . . . that it ends up like liturgical confetti.

This dialogue stuff (me Jew, you Christian) is all very well, but if you persevere you get beyond dialogue and actually touch God within whatever you're dialoguing with. It's like touching a live wire – you get an ecumenical electrical shock. It hasn't made me want to convert, but it certainly converts the way I relate to my own religion.

I make my own crude translation to make sense of what's going on around me. Some things are straightforward. Mary is no longer a blond Barbie doll, or rococo kitsch, but my yiddisher momme – like my granny in fact, swathed in black shawls with slits in her shoes for her bunions. And original sin means for me we all live in an imperfect world so forget about perfection (so destructive) but get on with it. And I don't believe in eternal hell, even for Hitler. Nor are miracles part of my world. But that's enough for the moment.

I also close my eyes, leap over the culture shock and read St Teresa. I'm not into Reconquest Spain with its triumphalism and its uncertainties. It's still overshadowed for me by '1492 and all that' (no not America, but the Jewish expulsion). Actually I like her. She's more Jewish matriarch than I expected – a sort of Golda Meir of the spirit – both tough cookies! Having been a religious bureaucrat I like competence.

Over the years there have been changes of course. No more being read to at meals, which I adored, or unyielding board beds which I didn't, though my osteopath might. Of course I'd prefer the *ancien régime* with wimples (cute) and disciplines (not so cute), and no central heating. But then Christians prefer their Jews with earlocks looking like extras from *Fiddler on the Roof*. Providing you don't have to live in it, religious fancy dress is fine. It's only when you take final vows and know the door is shutting behind you that you really understand – and a one-night-stand visitor like me doesn't get that.

Was my rather unnecessary journey really necessary? Well, there was a lot of human affection from the friars and retreatants there, and a feeling of family. Also in that silence, their companionship helped me get into another gear, step into another dimension, touch something, see the world inside out or whatever. A colleague of mine occasionally comes with me to escape the pressure of parish life. At school his little daughter was asked by her friend where her father was. 'He's gone to a Priory,' she said. 'What's a Priory?' asked her friend. 'I don't really know,' she replied, 'but I think it's a rest-home for rabbis.' I wonder what Teresa, John and Thérèse would have made of that!

But I have to stop, because I'm in a hurry and have to pack for Berlin . . . or being Wednesday would it be Cyprus? You often need a holiday after a holiday – lots of holidaymakers say so – why not try a retreat house, convent or monastery? You won't need more than two or three days.

LB

Thoughts and suggestions

Sometimes people think they have a vocation when what they need is a vacation. The stresses of modern life have got them down and they badly need a rest even more than the consolations of religion. A good retreat can supply both and it might be more relevant to your life than a Costa.

There are centres which combine a rest with holistic therapies. I have never tried them myself but friends have spoken well of them. I have been on retreats and they have helped me a lot. There is a National Retreat Association which publishes a magazine called *Retreats*. I should make contact.

Can you trust an inner voice? Are you the ventriloquist and it the dummy? Is it a sign of incipient schizophrenia? For a normal person the test I think is this, and it's the only one I apply to myself: Does the inner voice lead you into cloud-cuckoo-land, or into common sense with a more caring and compassionate concern with ordinary reality? Obviously a lot of our imagination gets into it and for that I would consult an experienced retreat leader.

You don't even need to go on a formal retreat. Give yourself ten minutes of quiet in a chapel when there is no service taking place. Don't run away from the quietness. Just let the outer world recede. It sounds simple but it's harder than it sounds and it can take some time before all the clamouring voices and memories in you die down. (I start worrying about the electricity or the gas – have I turned them off back home? Then I can get on to self-pity and then embarrassment. Don't fight them – just let them die away.)

If you find the very idea of meditation or contemplation too daunting, take a book with you and let your mind think about a sentence or a word you've read in it. It can act as a diving board.

Marcus Aurelius said you should make a temple inside yourself. Silence is very powerful, as both he and the Quakers testify.

LB

Language

A friend of mine has the ability to devour new languages. He goes on holiday with a Serbo-Croat grammar and comes back fluent, with maybe a local dialect thrown in for good measure. He's not Jewish but used to sit on the bus in New York reading a Yiddish newspaper.

I am pretty comfortable by now in French and German, and though my modern Hebrew is patchy, the classical language goes with the job as a rabbi. This is not to show off, but it must be a gift of some sort, plus a lot of work on grammar, vocabulary and practice over the years, the pain and boredom of which I have mercifully forgotten. Perhaps somewhere behind such an ability is the need to communicate. We can mime a lot to each other, and sometimes eyes can say more than a dictionary, but talking is what it is all about. And learning someone else's language shows a commitment to others that people respect.

I still get a childish thrill about being able to get by in another language, however clumsily. French was my first school language – 'Un, deux, trois, où est le roi?' 'One, two, three, where is the king?' Maybe our holidays in France as a child gave me an impetus. Though, my willingness to speak French was not always appreciated.

We must have been driving through Paris on the way to the South of France when my dad went through a red light and was pulled over by the police. 'I'll explain what happened,' I offered. He gave me a dirty look. 'We don't speak any French!' he snarled and turned on his Canadian charm and helplessness for the gendarme. It must have worked. I've also had to play the innocent, languageless tourist from time to time.

What is the effect of changing languages? A friend of mine from Holland, who has lived most of her adult life in England, once pointed out that her entire love life had been conducted in English and there were whole ranges of experience for which she had no vocabulary in her native language.

It is not only a matter of vocabulary. There is something about the intonation, diction, sound of another language that also affects us. German makes me more conscious of the structure of what I want to say – waiting to fit in all those verbs at the end. Churchill's 'that is something up with which I will not put' makes perfectly good German sense. Perhaps the language knocks a bit of spontaneity out of me.

In French I flow and soften. The words roll into one another as if the hard edges are knocked off them. Am I different? Or is this just a more refined way of being prejudiced? Certainly biblical Hebrew must have had an effect on the behaviour of the children of Israel. The verb is the main thing in the language; nouns are only verbs that have stopped moving, 'frozen'. And since the verb usually comes first in the sentence, before even the subject, biblical characters are off and running, always busy doing, and only thinking about it, if at all, afterwards.

All of which is just to point out how much more fun a

holiday can be if we can get our tongues round a few words of the local language. It turns us from tourists into visitors, from strangers passing through to potential friends who can be taken seriously. It opens doors as well into people's private worlds that would otherwise stay shut. Of course if all we want is two weeks with suntan oil and cheap booze at Bernie's taverna ('a touch of back home on the Costa'), it's not worth the bother.

But there may be more to the whole business. The biblical story of the Tower of Babel tries to explain why different languages came into the world. Once there was only one language which everyone spoke, but when they tried to build a tower to heaven, God came down and confused their languages. It's not clear why – God wanted people to explore the whole world and they wanted instead to stick together where they were. But whatever the reason, there is one odd nuance in the story that usually gets lost in the translation. When God scrambled their language it usually reads 'so that a man could not *understand* his neighbour'. That is one meaning of the Hebrew verb *sh'ma*, but not the usual one. It is mostly translated as 'hear', as in the famous 'Hear, O Israel, the Eternal our God, the Eternal is One'. So does it mean 'hear' or 'understand'? Perhaps the problem was not that they could not understand each other, but that they just weren't prepared to listen any more. They let the difference in languages get in the way of communicating with each other. If we want to understand each other we can, whatever the language we use. If we don't want to understand each other, we can do that perfectly well even if we speak the identical language.

So learning another language is a great spiritual challenge because we have the chance to meet new worlds and discover how different and how similar they are. Opening doors to others opens doors within us as well. So, next time you pack a little phrase book, or leaf through one at the airport, and wonder how often you will need the phrase 'Where is the nearest apothecary, my postillion has been struck by lightning', learn it anyway and give it a chance. The next 'apothecary' you meet might become a friend for life.

JM

Crossed-word puzzles

I can't help collecting those mangled English translations that turn up on signs in other countries. Like the one above the taps on a French train: 'Turn the lever indifferently to the left or right.' Heigh-ho!

On the German autobahn I found this notice aimed at making us feel welcome: 'Please leave the trays on the table. We want to make your stay in our house as comfortable as possible. You may leave the china in best conscience!'

Apparently a railway station in Hong Kong announces at regular intervals: 'Beware of your luggage!' Another one warns: 'Beware of announcements!'

A French hotel had a questionnaire about your stay that was translated into impeccable English. The heading to the nicely presented document read in large letters: 'Did you like it?' But the French set me off in quite another direction. I would have understood 'Avez-vous aimé' to mean: 'Have you loved?' Apart from the impertinence of such a question – what business is it of theirs? I was very impressed that they could immediately come to the existential centre of my being. I'm still trying to figure out the answer.

JM

Try another language of prayer

The prayers we say have been said so many times, many have become mere charms or mantras. Try saying them in the language of the country you are staying in! The result might surprise. I always dismissed the biblical injunction 'Be ye perfect!' because perfection is not a characteristic of our world. But in old Hebrew, there is no word for perfection. The words that it translates really mean 'whole', 'complete' or 'innocent'. Those made sense.

LB

Broadening my mind

Travel did broaden my mind whether I was in a pack of voracious youngsters youth hostelling or in a package to a hotel whose concrete still hadn't dried.

My earliest disordered journeys didn't give me much culture but they certainly prevented me from becoming a prig. Hitchhiking across Europe in the years immediately after the war was my university. There was no stigma to hitchhiking then. Cars were few, petrol was in short supply and the railways were only just recovering a sort of service. And people were travelling everywhere, criss-crossing the Continent, sorting out their lives in the fallout and debris of the war – and I listened in hostels and hedges and workers' restaurants to the stories of their lives. There were fascists making their way to South America and Jews to Israel or North America, and young people of my age without family looking for anywhere which would let them in and where a decent middle-class life was possible. There were Dutch youngsters fleeing the call-up to war in Indonesia, POWs running away from repatriation to the Eastern bloc and idealists running in the other direction to the Soviet Union to help Uncle Jo build utopia, God help them! And among them were all the curious youngsters like myself wandering to the Med.

longing for love, romance and dancing under the moonlight to the songs of Charles Trenet and Carlo Buti. My tight maxims of suburban life burst under the strain of comprehending such hopes, pains and fears.

My wanderlust never left me even years later, when I was a rabbinic student, trying to fit life into the corset of laws and theologies I was learning at college. They didn't fit. When you're a traveller, people can unburden themselves to you because you will never meet again. So railway stations and roadside buffets on my first holiday packages became confessionals, and what was said in them more convincing than the ritual confessions and ritual atonements I was learning at my seminary.

Sheltering from an unexpected blizzard in the crowded bar of a continental station, a Swiss lady asked me what I did and I told her I was training to become a rabbi. Since I was a minister, could she tell me her problem? Feel free I told her and put on my sympathetic listener's look. She'd been happily married for years she said and then her middle-aged husband wouldn't make love any more. He refused to go to a doctor or counsellor. It was all dirty, he said. On their last holiday she'd hit the bottle in despair. She had her need, what should she do? I didn't know or didn't care to say and was relieved when my train was called over the tannoy.

I couldn't or wouldn't help her but she helped me because I didn't give safe saccharine sermons on sex after that.

Other teachers who taught me my ministry were equally unexpected. Dutch old ladies were important to me. On a student package in Holland, I noticed the old ladies in my street didn't peer furtively behind lace curtains, British style, but ensconced themselves comfortably by large picture windows with apple cake and coffee to fortify them. Two angled mirrors were fixed on the outside of their windows so that they could see down one end of the street to the other. Their curiosity was blatant and unashamed.

But what intrigued me was that though they wanted to know, they didn't want to judge, and in that way they were unlike us in England where the media were getting more salacious, more moral and more judgmental day by day. Perhaps it was their Calvinism or just common sense in a crowded country, but Holland remains an unjudgmental country to this day and living in it taught me to be unjudgmental too. And this is religious though it may not seem so, for Jesus says, 'Don't judge lest you yourselves be judged.' And the Talmud adds, 'Don't judge people till you've come into their place.' Human beings are so complex, if you can, leave it to God! So no snap judgments on holiday either.

Over the years I've formed the habit of trying to read the literature of the country that I'm in. So when I visited Paris I bought a copy of the poems of François Villon from the bookstalls that line the Seine. He was a poet, a criminal and a Parisian in fifteenth-century France. In his last poems, written in prison on a charge of burglary with murder, he imagines himself as one of the skeletons of the condemned which used to hang just outside the city gates who talks to a smug passer-by, staring at his bones. 'Not all of us', he says, 'were born sane and sensible, so ask God to forgive us all, both you and me together.' These words have helped me distinguish real religion from hypocrisy.

I am so grateful for the wisdom I learnt on the hoof. I could never have learnt it in the security of my

English suburban home. You have to journey away from your domestic protection to be open to subversive truth.

It is very easy to forget the truths we have learnt on holiday. As soon as we are back so much of it seems like a dream. In former times people kept spiritual diaries to remember these truths. A diary may be too organised a venture but a small pocket notebook will help you not to waste the moments of revelation and the hints of higher things that were accorded you during your journey through countries and through life.

LB

Two questions to my teacher

The best religious teacher I ever had, who was overwhelmed by the violence and evil of his time, said this when I asked him what the last judgment would be like. He answered: 'God will take us one by one to Him, set us on His knee and then tell us what our life was really about. And we shall see truly the good that we have done and the bad and that will be our heaven and hell.' 'And what do you think He will say about me?' I couldn't help adding. My teacher thought and replied: 'He will say, Lionel, you have done just what I thought you would do – no better, no worse.'

He died not long after this conversation. He endured so much in the Hitler time, I think God has some questions to answer too.

I also remember other words of the poem of François Villon:

Broadening my mind

Fellow human beings who live after us,
don't harden your hearts against us.
If you take pity on us
perhaps God will take more pity on you.

Remember not all of us were born sane and sensible!

LB

Seeing ourselves as others see us

Holidaying on the Continent, I am invited to dinner by an English couple I scarcely know. I accept and am introduced to my fellow guests: another nice English couple, a Dutch one and a single French lady. Are they trying to pair me off with the French lady? Surely not! Over cocktails all becomes clear. They are all avid, cross-channel listeners to the BBC, on which I have recently spoken and some have read my books (bought or library?).

Aren't the presenters friendly? Yes. Was my little talk recorded? No. Who provides me with my Jewish jokes? Colleagues and taxi drivers, hospitals and congregations. Do I censor them? Yes. The English lot insist I tell them an uncensored one, they are so broadminded. I foolishly oblige with one from a taxi driver about two Jewish ladies and a bunch of flowers. It's anatomical but not immoral. The Dutch couple fall about with laughter, the French lady giggles, the English men give me an old-fashioned look and their wives look down their noses, which annoys me as they were the ones who insisted on it. Later I find one of them doubled up with laughter on the stairs.

Now English people are a good-natured lot, who don't mug or murder much, and I'm pleased my grandparents landed in London not Ellis Island. But only abroad do we spot our special English vices, which are hypocrisy and self-righteousness, denounced more by Jesus in the Gospels than by our British establishment.

That's why we got so angry with Brussels when we were leaders among the Mad Cow culprits. That's why we love peeking into the private lives of our politicians, which are none of our business, why we self-righteously and gleefully moralise over our royals and cover our gambling with a fig leaf of charity and good works. That's why we find it so hard to be straightforward about our snobbishness, sex life or euthanasia.

No, I won't finish the story about the two women and the bunch of flowers either in this book or on any programme. I know you're all broadminded but I've heard that one before. Here's one just as anatomical but less basic. An Englishman in a chic Paris restaurant points to a fly on his salad. 'Le mouche,' he exclaims austerely. 'La mouche, monsieur,' corrects the waiter. 'Good heavens,' says the Englishman, 'what wonderful eyesight you've got!'

A simple, tough exercise! It's not enough to spot the hypocrisy in other nationalities or even in your own. Can you spot it in *you*? I tried – it's jolly difficult. Give yourself a Brownie point for every instance you can remember and are prepared to admit to!

LB

The gnat – how holidays broaden the mind in unexpected ways

The ministry is not a highly paid profession and many clerics find it hard to socialise in a middle-class congregation. They haven't got the means. So like other low-paid professionals they try to combine work with pleasure and have a free holiday by leading a pilgrimage party or attending and lecturing at a conference. This busman's holiday doesn't work well with me where the essence of a holiday is a complete cut-off from all the ordinary demands and preoccupations of my job, even from the people I know and minister to (except God), though I am very fond of them.

But whenever I travel and move out of my environment my senses get sharpened and I notice things I would never have done at home. Does travel broaden the mind? With me certainly, though in very unpredictable ways. At one such ecumenical conference the predictable parts were pleasant and kind. Nice nuns explained their lives and vocations to Free Church people, and Evangelicals to Anglo-Catholics, while I explained to anybody who would listen what a Jew like me was doing among them. But a moment of unprogrammed enlightenment came to me from two creatures who didn't figure on the programme – a real gnat and a fictitious cow.

I noticed the gnat as I sat beside a lecturer and watched it trying to cross the table – though why, because there was nothing there for it on the other side? But then dogs also look as if they're trotting to urgent appointments, but they're so easily distracted by other dogs' bottoms you can't believe them.

This gnat seemed set on a pilgrimage of biblical proportions. It hesitated before a ravine between two piles of lecture notes. But then somebody asked a theological question, and the speaker shuffled his notes, nearly murdering my gnat who now faced a cliff not a ravine. Every valley had been exalted for the poor insect, just as Isaiah prophesied.

I decided to make the crooked straight for the poor thing but it nearly fell off in alarm. I would never understand its thoughts – it would never understand mine. How could I explain to it, for example, the theology that had nearly murdered it? And yet are we human beings so different from gnats, crawling on the surface of a speck in space whose vastness and purpose are beyond our understanding?

I mentioned this sad fact to a Dutch delegate in the bar who tried to cheer me up with Dutch gin and jokes. 'How does a cow hunt a hare?' he asked. 'Well how does it?' I replied obligingly. 'It hides behind a tree and pretends to smell like broccoli,' he answered.

Perhaps you get it – I don't, and it wasn't the gin because I'd heard that joke before in Holland from a politician and a publisher. How little we know not just of gnats but of our nearest neighbours in the EU.

Although I didn't learn much from the fictitious cow, later on I did learn a lot from real ones. Reluctantly I had allowed myself to be persuaded to go on a cheap, self-catering, country holiday. It was not a success. I am a town person and need people to chat to. But a friend of mine who does like the country advised me to chat to my fellow creatures, though not my fellow human beings – cows, real ones!

I thought he was having me on, but he was right. They do get interested in their muzzy way when a human being talks to them and they gather round on the other side of the fence to listen. Of course they don't understand the details of what you're telling them, but they seem genuinely sympathetic and on your side, which is more than you can say about a lot of your fellow humans. I found talking to them very therapeutic, which made my unenthusiastic country holiday worthwhile. For the first time in my life I tried to understand the way they looked at life as they were trying in their dim fashion to understand mine. Do they resent us taking their milk, I wondered? Wasn't it intended for their calves? Such questions would never have come into my mind back home in a London supermarket. Travel had undoubtedly broadened my mind which is what it should do because your mind, as well as your body and soul, has to be satisfied on holiday.

Therefore I advise you to *exercise* your mind on holiday. Use it to cross a barrier of prejudice and increase understanding. Converse with Germans and gays and Americans and Japanese with cameras. You wouldn't do it at home, but now is the opportunity. That is what travel is for, packaged travel, any kind of travel. Precisely because such learning is unprogrammed, we learn more.

Looking back on a half a century of holidays, it is the memory of conversations in trains, in hotel foyers and on dusty roads that have remained with me. They were my education. Risk it!

LB

A holiday occupation

Examine your small fellow creatures on a beach, in a park or on your balcony. Don't give way to your inclination to destroy them. They present a puzzle which may have a lot to tell you and on a holiday you have time to puzzle over it or puzzle it out.

In November 1785, a nest of mice was turned up by a plough in Scotland. Robert Burns learnt sad lessons of life from their fate.

But Mousie, thou art no alane
In proving foresight may be vain.
The best laid schemes o'mice and men
    Gang aft agley
And lea'e us nought but grief an' pain
    For promised joy!
Still thou art blest, compared wi' me,
The present only toucheth thee.
But och! I backward cast my e'e
    On prospects drear!
An forward, tho' I canna see,
    I guess an' fear.

LB

Happiness

Holidays are about happiness, which some of us never seem to get from them, and many of us not adequately. Which is why we so often lie about our holidays, prettying them up or falsifying them in retrospect, because we don't want to appear to ourselves or others as if we have failed.

Happiness is not easy to get in this life and even more difficult to hold. It is not an easy proposition. So don't blame yourself if all your hopes and fantasies aren't realised. They rarely are, whether they concern holidays, relationships or ambitions.

These are some metaphysical observations about happiness. If you absorb them they will save you a lot of holiday heartache, though they may not fit your mood as you clutch your tickets and brochures and all is *couleur de rose*.

The first lesson about happiness is not to run after it – it is too unpredictable and only runs after you. Only when you stand still can you both catch up and bump into each other.

The logic of it is peculiar. You only get it by giving or spreading it. The more you give the more you have. It is not like a cake where the bigger the slices you give away the less cake there is left for you. If you make other people happy on holiday you will probably wake up one morning and suddenly discover to your own surprise that you are happy yourself.

The main things you can give on holiday are your attention and consideration. Nobody ever gets enough of either. In this most people are givers not takers. Very few of us listen with all our attention or have an understanding heart. Don't worry if you are rejected. Other people are suspicious if someone else tries to help them – they've probably been rejected too many times to trust.

There's that chap by the bar getting sloshed and everybody is whispering tut tut. If he hasn't yet got sloshed, try to find out what's the matter. If he's an oldie he might have come back to a place he used to enjoy with his partner but is now bereaved or divorced. He might have been used to a lot of money, and now lives on a small pension. It helps if he can talk a bit about his problems or his partner. But do listen, don't pretend to listen, that's easily spotted and rightly disdained.

This need for attention applies to youngsters too. Being stood up is part of the ruthless sexual chase and you can cast your mind back and remember how hurtful it once was to you.

Also, be considerate to the hotel staff. That old lady is waiting for you before she can clear up and get back to her family. And don't be mean with tips, though don't buy people.

This sounds unpromising stuff for happiness. But it works this way. If you do something for the sake of heaven, heaven happens and whenever heaven happens a release of happiness comes with it.

Some people think in a very different direction. If only

I had more money and power, they say, then I'd be happy. That is true if they are really suffering from oppression or grinding poverty. But most holidaymakers are not in that class.

LB

Thoughts

Here are some thoughts on happiness that you may find useful.

- For happiness stand still and enjoy what you've got – what *you* have got and not what the *other* person has got! The latter is the way to perpetual unhappiness fed by jealousy and envy.
- Think back to the happy times of your life. What gave you your happiness? It's no use floundering around in life without knowing what it is you actually want.
- Will you still want it when you've got it? Lots of people, for example, sigh for a partner, a wife, a husband, but run away as soon as a possibility comes along. Are you more comfortable with your dream than with reality?
- Don't be a doormat!
- Try not to push!
- Don't seek perfection! It's not something the world can give.
- This saying comes from Solomon Ibn Gabirol who lived in Spain in the eleventh century. It also turned up in a pop song. 'If you can't have what you desire, desire what you have. Don't live in perpetual gloom.'
- Perhaps happiness is easier than we think. Krishnamurti advised people who wanted to be happy,

'Be happy then in whatever you're doing!' He also told us not to be imprisoned by memories.

LB

Adventure holiday

I look with envy at the adverts and convince myself that one day I might follow through. With a camel across the Sahara. Two weeks up the Amazon. Through wildest Borneo on a bicycle. The call of the wild. The promise of adventure. The challenge of testing urban man against nature red in tooth and claw. Oh, the thrill! the glamour! the discomfort!

As I turn the page to the package holidays to Ibiza I give a sigh for the me that might have been and pass quickly on.

Of course, there were earlier times. Not exactly the rough guide to exotic lands, but I've stumbled into a few man-made adventures and dangers. There was that trip to Israel in the spring of 1967. War signs were on the horizon and my parents were anxious. In the event we landed en route at Athens airport only to be held up in the lounge for twenty-four hours during the revolution of the Generals. It was a good introduction to Israel on the brink of the Six-Day War.

I understand that war differently now, but at the time it seemed crystal clear who were the goodies and who the baddies. When I decided to stay on in Israel despite the circular letter from Her Majesty's Government suggesting that I take urgent steps to vacate the country, I went to

Hadassah Hospital and offered them my newly registered medical skills. They looked at my one-year's experience and put me in the delivery room – so I spent the war doing episiotomy repairs. Literally a busman's holiday.

There was one adventurous moment. I had been staying with another doctor near the hospital. The day after the war began we decided to return to the flat and pack enough clothes to see us through the next period as we camped out in the hospital. The car park contained some burnt-out cars hit by shells, and some of the walls were pockmarked by shrapnel. At that time the only news we heard on the radio was negative, part of a propaganda exercise. We expected the worse.

So we made a dramatic run between the buildings – I doubt it was necessary, nothing sinister was happening, but we entered into the spirit of the thing.

The Hadassah Hospital is at the top of a small mountain or large hill depending on how you define these things, with a winding road leading up to it. As we drove, my medical friend, in what could only be described as British cool, announced that he had been in the cadet force at school and we should look out for passing planes and point them out to him. He would then 'spot' them and tell us if they were friendly or enemy ones! By then I was beginning to think we were overdoing the dramatics. So I asked, unkindly, what evasive action he intended to take in the event of 'spotting' an enemy. We could either drive into the side of the cliff or off the edge, neither of which seemed very helpful. I remember he got angry because I wasn't taking this business seriously enough. Still, we got home safely and I survived the war, at least physically unscathed.

But this in turn reminded me of my own cadet force

days. Having flat feet I managed to get transferred to the Air Force section – far less marching and the chance to fly the school glider. (It was towed by a long bit of elastic and usually rose at least three feet into the air.) But one extra adventure was a trip to some military camp for a weekend. We were sent off on a 'night exercise', a test of our endurance and initiative. We were dropped five miles outside the camp with a map and compass and told to get back through 'enemy lines'. Our corporal, like my medical friend, took it terribly seriously. I suggested that we walk back along the road we had just taken by coach, but he struck off across country like some ancient Moses heading for the promised land. In the pitch black, stumbling through fields, we got hopelessly lost. Finally on the horizon we saw a building. It was an isolated farmhouse. By now it was two in the morning and no sign of life around. So we tossed pebbles against a window and then called out. A light went on and an elderly woman stuck her head out of an upper window and asked what we wanted. There are such moments when you only get one question and it has to be the right one. Our corporal rose to the occasion and shouted to the bewildered lady: 'Which way's north?' It wasn't the right question, the head disappeared and the window slammed shut.

We staggered on, tired, cold and cowed. We hit a road and I spotted a bus stop. Asserting my natural leadership qualities I suggested that we waited there for the bus since it would take us straight back to the camp, and eventually we rode back on the top deck. We thought we had done pretty well, but were told we'd been spotted and shot by the defenders. Next time I'll duck down.

I shouldn't be too proud about my inability to manipulate the natural environment. Some people take very seriously the possible collapse of 'civilisation as we know it' and are preparing to rough it in some kind of post-nuclear-holocaust wilderness, living off mutated plants and creepy crawlies, and defending their patch of territory to the death. So maybe one day I will pack my rucksack, take my shots, go on twenty-mile hikes to prepare the ground and head off into unknown, untamed territory. Meanwhile I will settle for the wilderness in my head.

But is that enough? You don't have to go into a wilderness to have a spiritual adventure. In fact any journey into unknown territory, from a day trip to the seaside to a package holiday in Florida, is a kind of test. Because a new place, new sights, new people offer a chance to expand our own experience, if we are brave enough to follow through. We get caught between two desires. The first is to explore all the new possibilities before us. But the other is to tame it, to find some way of making it familiar and safe, as near as possible to what we have at home. So do we try new foods or look for fish and chips? Do we try to learn a few words of another language or speak loudly and slowly in English? Excursions can certainly take us to somewhere new, but do we also head off the beaten track to find out for ourselves (having checked out the risk of bandits or kidnappers beforehand!)? So every holiday is a chance to stretch our minds, our hearts and our souls, if we are willing to risk at least one completely unfamiliar experience. So 'boldly go' on at least one journey to where you have never gone before. And when you get home it too will be different because of what you have discovered about ourselves and the world outside.

JM

Thoughts

> Who is wise? Whoever learns from everyone. (*Sayings of the Fathers* 4:1)

> God said to Abram: Go, for your own sake, from your land, from your birth place and from the house of your father to a land that I shall show you. (Genesis 12:1)

There is a story about a poor man in a little village who had a dream that if he went to the great city of Prague he would find a treasure buried under the bridge. Since the dream came three times he knew it must be true and so he made the long journey to Prague. But when he found the bridge it was guarded, so he visited it every day hoping to find a chance to look beneath for the treasure. He began to talk to the guard who learnt his story and began to laugh. 'If I believed in dreams I would go off to some little village', and he named the place where the man lived, 'and dig for a treasure under the bed of "so and so" ' – and he named him! So the man thanked him, returned home and indeed found the treasure under his bed. With it he built a place for religious studies.

Sometimes we need to go on a journey to discover the value of what we have back home. Sometimes a chance encounter with another person can help us understand more about ourselves. But first we have to move.

JM

Spirituality in the saddle

Before I go on an inner journey I have to take an outer one, preferably free-range not packaged, which is why I took a cycling holiday in Holland, where I learnt important spiritual lessons in the saddle.

Now, I'm not built for bicycles – I'm too plump and not sinewy enough – so I had to watch little old ladies with mighty Dutch thighs smile and sail past me. Which was good for my vanity. Also, I haven't got a good sense of balance, so when I gave a hand signal, I fell flat on my face in front of the car behind. 'What did you do that for?' shouted one shaken motorist. 'To be helpful,' I said indignantly. 'O God,' he pleaded, 'don't be so unselfish in the future.' Which is curious advice to give a minister of religion.

Each night I stayed with families who opened their farms and homes to genuine cyclists and walkers. They charged about ten pounds for bed and breakfast. My bed was spotless and the breakfast table was loaded with cheese, chocolate and fresh eggs. Some told me to make sandwiches out of what was left for lunch. They were brave ladies and I'm not sure I'd have been so welcoming to three grubby pensioners in the dark, claiming to have lost their way in a turnip field.

I'm also indebted to those ladies for an unexpected lesson in business which I badly needed. On the way to Holland I'd read in a paper that only business could save Britain. But business had got so sleazy I was doubtful if it could even save itself. Some bureaucrats and old boys were awarding themselves outsize perks – legal but unlovely. Some salary and wage demands concealed the same 'gimme gimme' under a fig leaf of fake concern. Some of the not-so-genuine, as well as the genuine, got money out of people using dogs, babies and blankets on street corners. And gambling is now a national way of life because everyone wants a fast unearned buck.

Now I'm not naive. I know the Dutch B & B business is no big deal economically, but it's still an example of commerce that could be both decent and efficient. And because of it, I'm just that little bit more hopeful.

I also learnt an important lesson in contentment in Holland from an old Amsterdam story about Sam and Mose. Sam sees Mose walking in agony, wearing shoes several sizes too small for him. 'Why don't you get the proper size?' he said. 'Look,' said Mose, 'I've been ill, and business is so awful I sometimes think I'll make an end of it. But when I get home and take off these shoes, life seems so wonderful I bless God for it!'

Which was the same with me when, after my holiday ended, I got back and sank into my high-backed, padded office chair, no longer riding a leather saddle over cobblestones. Then I thanked God for my holiday which had been great and I added from the bottom of my heart, 'Lord, isn't ordinary life lovely too!'

LB

A thought

Holland is a little country and you learn little practical bits of spirituality in it. But these humble lessons are the backbone of life. Never despise small things! When both the rabbis and the gospel-writers talk about the kingdom of heaven, their ways into it are a lost penny, someone who goes on a journey, and a light behind a bushel. If you have no book of profound spirituality on you, meditate on the little things that happen to you in daily life.

LB

Spiritual walkabout

My friend Evie and I got hooked on spirituality and every so often, disillusioned with London life and the ecclesiastical rat race, went on spiritual walkabout round the country, calling in on surprised monasteries, convents and retreat houses to graze in fresh green spiritual pastures. Were we Christian? No. Were we married? No. Were we . . .? No, we weren't that either. Most of them took us in and, as they had enormous buildings with few novices, allotted us floor space for our spiritual quest.

Occasionally we caused panic. 'My God, 'tis a woman,' you could hear some think and, after much whispering, I would be assigned a cellar at one end of the monastery and Evie a turret at the other – which was kind as they were old and we were odd and disturbing.

At teatime, all mumbles ceased ominously as Evie entered and sat down at the all-male refectory table. 'May I have some bread and butter?' she asked. 'No,' croaked an ancient, and the other ancients cackled and waited to see whether they had driven her away. But Evie continued her tea obstinately in the hostile silence.

After meditating in the chapel I found Evie had pinched my aftershave. She was having a ball. 'O you poor dears,' she cried as the wrinklies chatted round her while she listened attentively and brushed and clipped their beards, removing egg stains from their dirty habits, polishing their nails pink with her manicure set, and sprinkling them with my aftershave. They purred with pleasure. At her insistence I told them about a teacher who asked his class who was the first woman in the world. 'I'll give you a clue,' said the teacher, 'it's an apple.' 'Granny Smith,' shouted one boy proudly. The ancients fell about with laughter. Though properly watered and fed, no woman had cared for them lovingly, and unloved old men get surly and unkempt just as unloved old women get finickety and spikey.

Evie couldn't make our next spiritual walkabout. She was too preoccupied with a sheep she had bumped into in a high street after midnight. Was it a pet or would she have to save it from sacrifice or the pot, perhaps I could help her? Life with her was like the perils of Pauline.

But before she died I had learnt from our last spiritual walkabout that the spirituality we seek is already inside us, otherwise we wouldn't want to make such a trip in the first place – the people we meet only trigger it off. Also she had retaught me a truth nearly forgotten in the new fashionable fanaticism with its hunger for religious results, that religion, whatever its theology, sociology, buildings, texts or titles, has to be validated by acts of loving kindness otherwise it is hollow. Such acts are its heart.

LB

Two thoughts

If someone comes to you for help, don't say, 'God will help you.' That's being disloyal to God. God sent you to help those who need you, not refer them back to Him! (Chasidic)

Being needed can be a pleasure and an honour. Most of us need to be needed.

LB

Places

Hotel room

The cautious books tell you to examine your room before you accept occupancy, but that is a council of perfection, not package tours. It's like the phrase books in which you command the waiter to bring you your uncooked fish for inspection. (What! On an all-inclusive budget!)

Still, it is advisable to insist on a change of room straightaway if you really can't stand it. If giant lorries are fuming and honking away beneath your balcony, or your bed is just above the organ pipes, hotfoot it back to reception and don't be a shrinking violet or a doormat. Sometimes the hotel is really full, quite often it's a try on.

Your room will be basic, not perfect unless you have splurged on five-star. That's reasonable because the hotel management can't do everything. It's up to you to turn it into home. How do you do that?

Don't charge down to the bar as if you were being shot from a cannon and rejecting your room. Sit a while and examine it. Start to make friends with it. You can buy some flowers and make a vase out of a cut-down plastic water or cola bottle. That brings a bit of enchantment. Even plastic blooms will do. You can sprinkle them with aftershave or Eau de Something for a sophisticated summery effect.

If you're religious, or just superstitious, say a prayer for all the bods who've slept in your bed before you, and will sleep in it after you, and then invite God in or His angels. If this seems too fanciful, just say a thank-you to Whomever or Whatever for the peace of the place.

Now plump yourself down on the bed or best chair and make it your home by eating and drinking something nice which you've reserved for the occasion – a miniature bottle and a chocolate you've kept back from the plane journey would be fine. Look at your room in different ways. How would Vermeer have painted it, or Dufy or even Dali – though on reflection, better not that one?

Here's a comfortable spiritual exercise. Sit back and think of all the nice things you've done over the last few days – the bores you put up with, the coins you gave beggars, the tempers and tantrums you didn't give way to, though packing and getting away was a stressed time. Begin to like yourself! If you like yourself, other people will sense it and begin to like you. Forgive yourself your trespasses and forget as well as forgive the trespasses true or imagined against you. Give yourself a holiday from the bully that lives inside you.

If you're not schizoid, you can ask God to sit on the stool or another chair (as it were) and start to chat with Him/Her/It. Tell Him how you really feel. Listen to any comments and answers that form in your mind. But don't force them.

If you feel a bit weepy, indulge. You've got to let the stress out some time. That's what your room is for – privacy!

Also, keep a spiritual book especially for your holiday – a user-friendly one not a tome. Just read a few lines each night. This will satisfy your soul which isn't used to much. The ones I used to take were *A Year of Grace*, published by Gollancz, or *Becoming What I Am* by Harry Williams, or a children's version of *Pilgrim's Progress*. They stop this world getting out of hand.

Remember, if you find your room hateful like a cell, it isn't really like that. It's just an ordinary hotel room. The hate is inside you and now you're on holiday it's coming out (which is good) and bouncing back to you off the walls. They act like reflectors or screens, reflecting back the feelings you project onto them. They will change as you begin to relax.

Here's another spiritual exercise. Turn a part of your room into an unpretentious private chapel, like many Hindus do, where you can find your divine self, not just your social self. And where you can enjoy yourself by being by yourself for at least ten minutes each day.

If you need to focus your attention, a dressing-table works very well as an altar or an ark. Your pew can be the most comfortable chair or cushion, and beside the flowers I mentioned before add anything which brings out the love in you – like an icon or a picture. Don't use candles for eternal lights because they set fire to things. Use a torch instead, but really the changing lights through the windows or balcony should be interesting enough.

Chapels don't need marble and altars don't need gold or lace antimacassars. Remember all these outer things are pointers to the more important altar and chapel which are inside you.

Explore your own way to rest and stillness. I like to lie flat on the floor with my head on a cushion. Light classicals on the radio help. Shake one hand first, then the other hand, then an arm, then the other arm, then a leg etc., till your whole body is shaking freely. It shakes out the dismals that get locked into your muscles.

It takes me about ten minutes to become quiet because I have to go through so many layers of anxiety and bad memories. When I have reached a kind of peace, I let my mind wander to all the people I'm going to meet. I think of the tangles they're in and my own tangles. When they seem less tragic and more funny, I say thank you to Whomever or Whatever, get up and go down to meet them feeling much chirpier.

LB

A suggestion and a thought

Looking at a picture or the reproduction of one can bring inner peace. I have used an interior of Vermeer (ordinary National Gallery postcard) and another reproduction of Gwen John's picture of her room. She was Augustus John's sister, and some think just as great. They go well with a pot of flowers. (You can give the pot away as a souvenir when you return home.)

> Without going out of your room
> you can understand the world.
> Without peering out of the window
> you can discern the Way of Heaven.
> (Lao-Tse – *c.* sixth century BC)

LB

My kingdom for a loo

Most of the things you need to know when you go abroad are pretty easy to find out, especially if you're on a package tour. But there is one vital thing that, for some reason of delicacy, never seems to be included in the brochures they hand you – how do you find a loo in the particular country you are visiting. Are they marked in a certain way? Are there public ones, or do you have to find a restaurant or other such place? Do they charge? And anyway what do you pay those mysterious men or women who sit in the entrance with their saucer filled with miscellaneous coins? Do they live off these tips or are they simply a bonus? And if you have to pay, do you do so on going in or on coming out? (In France once I was about to crouch over the 'hole' when the guy on the door operated the flush and flooded the cubicle. I won the standing long-jump record at that moment and saved my trousers from drowning – but I never found out if it was chance or malice on his part.) There used to be a 'good loo guide' in England; an international one, with tips for the unwary, is long overdue.

I remember my father's look of disbelief on encountering for the first time the old French 'hole in the ground', with its two impossibly placed raised footprints to stand on. There he drew the line. And I still recall the queue of irate, sun-tanned women standing outside the ladies loo on a beach in Cannes when my father eventually emerged, satisfaction on his face, indifferent to the cries of anger, happily puffing a large Havana cigar. Happiness is a loo with a seat.

Not being a smoker I've never dared to duplicate his feat. And in a way it would be a cruel trick. If there is any single design fault in human anatomy that can be hurled at the Creator, or at least at those who design toilet facilities for women in public places, it is the one which makes it impossible to pee into a small waist-high bowl in public. It is the one situation where all men are equal and all women have to queue, since there are never enough cabinets to accommodate them.

Men, too, have some inconveniences. There is a story of two gentlemen standing next to each other in one such stall in a European hotel. One starts up a conversation.

'Tell me, by any chance, are you Jewish?'

'Yes,' answers the other, a little unsure about what is coming next. The reason why he may have noticed seems obvious, but why mention it?

'So am I,' says the first one, as if to relieve his concerns. 'But tell me,' he continues, 'by any chance do you come from the village of X?'

'Why yes!' replies the other, intrigued and pleased at the same time.

'And, when you were a child, was the *mohel* [the man who performs circumcisions] old Reb Hayyim?'

'Why, yes indeed,' comes the reply. 'But what on earth made you think of him?'

'Well,' comes the answer, 'he had a tendency when

operating to put a slight twist on the knife – and you're peeing onto my trousers!'

A lady pastor friend is doubly affected by the problem of finding a loo as she often travels with church groups. Since the clergy are supposed to know everything about everything, temporal and spiritual, the moment they disembark from the coach all eyes turn to her in anticipation or pleading depending on the distance between stops. She claims to have developed a sixth sense and her eyes go immediately to the rest room – the reputation of the church once again redeemed, the spirit again overcoming frail flesh. Nowhere does it say that being religious should make you less competent in practicalities.

Which raises interesting questions about the provision of facilities in places of worship. I recall a marvellous throne in the Dean's house of Westminster Abbey. Though the nearest I've come to a religious experience in this regard is sitting on Freud's loo in Vienna, Berggasse 19. Did he sit here pondering anal fixation? I suppose one could actually organise a tour of the loos of famous people. Even the Bible that is usually reticent about such things, using euphemisms like 'uncovering the legs', goes to town in the case of King Eglon who was assassinated on the loo by a cunning left-handed judge called Ehud (Judges 3). Since he was accustomed to take his time it was a while before his servants were sufficiently concerned to overcome their embarrassment and enter to discover the body. If archaeologists could ever uncover his toilet a good entrepreneur could make a fortune.

This is actually no laughing matter. There is a Jewish prayer to be recited on going to the loo in the morning that thanks God for the many openings within us and reminds us what would happen if one of them was blocked. It is typically a matter we take for granted till something goes wrong – or the absence of tourist information leaves us hopping up and down in some distant street. The countryside is, of course, better, and many will attest that there is some corner of a foreign field that is for ever England.

Actually the loo is a great place for spiritual exercises. Where else are you left completely alone to meditate with no one to disturb you? Once in a while take the time to think about the marvellous way your body functions. Of course, if something is not quite right, one of the various forms of holiday tummy, your prayers may never be more heartfelt. But in such circumstances it never hurts to think about your own mortality, or at least remember how easy it is to take being healthy for granted.

Do you keep reading matter in your loo at home? If so, figure out which of your favourite books to take with you – or the ones you always meant to read but never got round to. The smallest room can sometimes be the place for the greatest thoughts.

JM

A thought

One of the Jewish law codes forbids thinking about religion and God in the toilet as it would be disrespectful. Instead you should think about business. But what do you do on the Sabbath when you are not supposed to think about business? Think about pleasant experiences. So the following piece deserves a more respectful place:

It is very desirable that we have a private room where we may study and pray in solitude. We certainly need such a room for self-isolation and the achievement of joy in our relationship to our Maker. Even sitting in such a room without engaging in study or prayer is desirable. If we cannot arrange things in this way, we can achieve such isolation when we are wrapped in our prayer-shawl. For the prayer-shawl may be considered a private room, particularly when we draw it over the eyes. We may even sit at an open book and while people may think that we are studying, in reality we may be communing with God. Thus there are many devices for attaining that solitude which is the source and root of holiness. (Nachman of Bratslav)

The Jewish blessing before going to the loo

Blessed are You, our living God, Sovereign of the universe, who fashioned the human body with wisdom and created within it many openings and many cavities. It is obvious and known before Your Throne of Glory that if one of them was opened or one of them was blocked it would be impossible to survive and stand before You. Blessed are You, God, who heals all flesh and acts in wondrous ways.

‘The last time I saw Paris . . .’

The first time I saw Paris on holiday must have been in the late 1940s just after the war ended and I broke free of my parents’ holidays. I loathed the London suburbia which I inhabited, constricted lives in constricted rooms, all furnished alike with a three-piece suite, a lamp standard and a box utility wireless. The de luxe version also had a phony glass cocktail cabinet. It lit up but you didn’t, because of course it didn’t contain any cocktails – only tonic wine. Instead I haunted Studio One in Oxford Street, smoked Players Weights while I watched French films and tweaked bra straps in the dark. In those pre-pill days that was the nearest we ever got to IT.

On the last day of school term I told my astonished parents I wasn’t joining them in Margate again ever, packed an enormous ex-army rucksack, and after lessons ended, joined the army of other youngsters hitching along the roads leading south from London towards Life (as she wasn’t lived over here), the Continent, adventure, Arletty (or Jean Marais), plonk, nights of passion (or days – I wasn’t particular), and the daring ones even talked of whores and horse (horse steaks not Longchamps – Britain still had meat rationing).

I myself lived on baguettes, cheap overripe tomatoes

and Gauloises, and struck up an alliance with a hard-boiled lass who wanted to hit the Riviera big. She carried the ubiquitous pepper pot to defend her honour against me and all comers.

So we journeyed jerkily through rural France. While she showed a bit of thigh along the road, I lurked behind a tree and when anything stopped, I staggered out with two bulging rucksacks. Some took one look at me streaked with tomato stains and accelerated. A few, because they were slap-happy, bored or holy, took us aboard. We got a lift on a hearse, on a gravel lorry, which left me streaked with blood as well as tomatoes (she shared the cabin with the driver, protected by her pepper pot), and in luxurious chrome-covered cars of Belgian profiteers, each with its own breathy blonde. The last introduced me to chips with mayonnaise, a sophistication undreamt of in Hendon, which I boasted about in my letters home.

My partner ditched me while I was lurking in a ditch near Paris, so I had to deal with the city of light alone, which I couldn't. I just didn't know where to begin. In London if you felt miserable it didn't matter because everybody else looked miserable too. But in Paris you felt a thousand times worse because everybody else seemed to be having such a good time, enjoying little causeries with lovers and mistresses (as I surmised from Studio One) over potions of passion in outdoor cafés.

There was certainly a lot of passion in Paris as the films said, but none of it was for me, and I cried myself to sleep deciphering Proust, while squeals of it came from next door and grunts of it from below. Which is how I first stumbled on Catholicism and the Divine, which shows God moves in very mysterious ways indeed, if it was God that is.

In a cheap left bank lodging (you shared beds as well as rooms), two young Irish boys threatened to beat me up, as I was English, Jewish and left, a combination they'd been warned against in the Gaeltocht. We compromised over steak (horse?) and *frites*. I would accompany them to their church and give them five Gauloises, and they would swear blood brotherhood (you did it with a pen-knife). I was willing – since human love hadn't worked I was quite prepared to give the divine sort a whirl, especially as I didn't believe in it a bit, being a Marxist.

But they got discomfited, not me. Getting high on the incense and Gallic fervour, I decided to join in everything. It was only polite and I'd promised. So I lined up with the others and received what I thought was Jewish Kiddush (bread to represent the necessities of life, and wine for the sabbath). This sent my blood brothers into a tail spin, and while they hurried off to find a confessional, I sat bewildered wondering what I'd done wrong. You just couldn't do right with Christians, as my grandmother said. Though the Irish lads kept clear of me after that and soon headed towards Rome, I continued to discover a peculiar Paris I'd never thought of before, of crypts, tombs, saints, and the first worker-priests.

I returned to Paris on holiday a few weeks ago. No longer looking like Rumpelstiltskin or Quasimodo under my rucksack, I slid effortlessly from London on Eurostar, consuming smoked salmon beigels and Bordeaux with a vintage. It was a smoother city than the one I knew after the war. But some things remained the same. The franc had risen and the pound fallen yet again, so fifty years on I was still sharing metres of baguette in the park with the pigeons. And that *bêtise* in the church must have made its mark, because while my friends went to the Baubourg and Orsay and the glass Pyramid, and Picasso, I was drawn instead to old St Severin, and a curious little crypt in Montmartre. To a synagogue where Aimé Pallière, a priest who loved Judaism, dared to take services after the occupation (he was buried with Catholic and Jewish rites), and to Sacré Coeur where Max Jacob, the Jewish surrealist poet, had a religious experience. (He died in the camps.) I tried to find a worker-priest but couldn't.

Now, as this is a spiritual book, I ought to sum up this slightly sordid but saintly reminiscence with a moral. I think it can only be this – God works in mysterious ways, which is trite but true.

LB

A further thought

If the divine draws you or just offers itself to you, or seems to wait for you, accept the invitation however and whenever it comes. Perhaps a guide is volubly conducting you round a chapel and suddenly you've had enough information. You just want to fulfil the purpose that chapel was created for and pray. It doesn't matter if you haven't prayed since Sunday school and have long forgotten the in-words and liturgical language. Sit yourself down comfortably and don't say anything, just let Whoever, Whatever come to you. He or She can come in a feeling, a word that doesn't want to let you go or just in quietness without anything more. If nothing comes don't worry or write the incident off too soon. It takes time for the mind to digest soul stuff. From my own experience I don't think that little visit will go out of your mind for good.

Don't worry if you only go into a chapel to rest your feet! We all go into religion for the worldly thing – but it eventually purifies itself.

Don't worry if you don't believe in the religion it represents! Why should you? There's a lot of junk as well as treasure in an ancient creed.

Don't worry if there are only a few black-shawled crones in it! (No, you won't turn into a black-shawled crone yourself.) Risk it! You're on holiday and your soul needs a bit of servicing as well. Explore!

LB

A secular city!

They were advertising bargain trips to Amsterdam. Once there, everything's on offer. It's such a secular city! You can buy bulbs in the flower market and nearly-new Rembrandts in the flea market. You can inspect the ladies in the red-light district and gentlemen's gentlemen in the gay one. Just by the Van Goghs, they put dollops of golden mayonnaise on your chips. Culture, herrings, eels or whatever, especially whatever, are there for the paying. Enjoy! Enjoy!

I used to gape at all that freedom, and then tripped up over what brochures don't bother to warn you against – God! There's a lot of sacred in a secular city if you don't reject its unexpected packaging.

Let's begin with the name. Only foreigners call it Amsterdam. To residents it's the Mokum, rhymed with hokem, which is the double-Dutch pronunciation of the Hebrew Makom. That's because before the last war, in another age, Amsterdam was a very Jewish city, the Jerusalem of the West, long before it became the Greenwich Village of the East, the Makom, the Place. For, as the rabbis said, 'God is the place of the world though the world is not His place.' You can figure out the metaphysics of that in a brown café over your first kopstoot, your headbanger, of beer and gin.

Some years ago when I was going through a religious dry patch – I had been a member of the establishment so long – I bumped into two 'oldest swingers in town' in a brown café. They were the last people I wanted to meet because I remembered how they used to feed me kopstoots in my student days to laugh me out of my vocation. Those guys were sweet but secular! But now I goggled at them. They were into purple pants and portraits of a smiling guru, and beads dimpled benignly out of forests of blond curled chest hair. They'd just come back from India they told me in unison, to confer with their spiritual director on the state of their souls, God wot! Well He certainly wots more than me, because they then apologised for not buying me more kopstoots as it was almost meditation time.

'But could I tell them about the Cabbala?' they asked. I inclined my head suspiciously and enlighted them accordingly. Two kopstoots later I discoursed on St Teresa and St Thérèse while they listened respectfully. I had never believed such things could happen to that pair of players. But God had got at them too via Air India not El Al.

I had preached to them once before I remembered. It was at a party where I made an awful faux pas. 'Why did the owner (whose name I never knew) have so many green straggly pot plants?' I asked. I wanted to buy him something more cheerful like geraniums. The guests and host, if he existed, burst into hip-happy laughter and retailed my remarks to latecomers, who snickered over this witty Englishman in their midst. How could I know they were all growing their own hash? The Dutch are frugal folk and it was only natural to combine house and garden with short-cut mysticism.

After some guitars got going, I was asked to entertain the guests with a sermon. I refused, aggrieved because I thought they were taking the mickey out of me again. But it wasn't so and the hip-hap hippies listened with much more attention than my normal congregation. Once again I had underestimated both them and the attraction of the divine. And most of them weren't stoned because I checked it out.

All ministers of religion get laughed at in Amsterdam. That's the Jewish inheritance of the place and it does us all a power of good, provided we don't take it too heavy. I've endured a lot of ribbing as a rabbi and the Calvinists and Catholics get it too as I heard from a Christian minister. Now in Amsterdam folklore, he told me, there's a puckish, Yiddish little character called Mose, much beloved of the population. Well Mose goes to Rome and His Holiness is delighted. 'How nice to see you Mose,' he says, 'we'll stroll on the balcony first and then we'll take some tea.' When they appear on the balcony the crowd in St Peter's Square below go berserk, shouting, cheering and waving. 'Who's that?' exclaims a bewildered Japanese tourist with his camcorder at the ready. 'Don't you know?' shouts his startled neighbour. 'Why that's Mose, Mose from Amsterdam, but I don't know who the other one is!'

But back to serious religion. Amsterdam religious monuments exist but they're not showy. The most moving Catholic church is not the ornate Cathedral but the church some businessmen built in their adjoining attics – 'Our beloved Lord in the Attic'. And in the Westerkerk there's a modest plaque which says that Rembrandt van Rijn was buried nearby in an unknown pauper's grave. I sit in the church and pray. I can't help it. And on a busy main road there's the façade of the Dutch Theatre

(De Hollandsche Schouwburg). The entrance is free, for inside is still a ruin. That's where they herded the deportees before their hopeless journey to the gas chambers. A small plaque says 'their sufferings were indescribable'. I pray again and cry. Again I can't help it.

There's another theatre, the Municipal Schouwburg in the Leidseplein. There might be a play by Vondel showing. He is to Dutch literature what Shakespeare and Milton are to ours. He was a friend of Rembrandt and a Calvinist who became a Catholic. He wrote the most moving Christian odes ever – also a plea for religious tolerance which makes me proud to be his fellow human being and his cousin in faith. He was one of the first to see God at work in a pluralist society, and I think they ought to consider sanctifying him too.

I arrived angry from seeing so much secular in the sacred (it's a problem for all of us professionally religious) but Amsterdam showed me how much sacred there is in the secular if I wasn't too snobbish to see it.

For a spiritual exercise read Ann Frank or Etty Hillesum – they are the unassuming spiritual triumphs of the Holocaust, people who started off much like ourselves but made something out of the tragic mess.

Notice the unassuming good deeds done by secular people in your hotel.

LB

Abducted by aliens

I thought I had stumbled into my own personal episode of *X-Files*: 'Tourist abducted by aliens in broad daylight – scientists are baffled!'

One moment I'm sitting in a bus being driven through the streets of Essen in Germany. Then I must have dozed off for a few seconds and awoke to find myself no longer in a bus but in a train going through a tunnel. We were definitely underground and on rails, not wheels. When I got over the shock I checked out my bus. It was more or less as I remembered it. But how far do you notice the inside of a bus, even in a strange town? All right, a first glance around, negotiate the alien (that word again) ticket system, find a seat if you're lucky, and then it's pure scenery – or pure anxiety if you don't know exactly where to get off. There may be other passengers to look at – and that might provide a useful clue if you think you've just been abducted by aliens. Does anyone else seem different or particularly worried by the experience? Then I realise I can't remember them either. Is this significant? Have they been replaced with pods? Have I been brainwashed as well? Or could it be that I didn't pay much attention to them to start with? Staring at them now seems to make them uneasy – but I'd become uneasy if I'd been stared at by some obvious foreigner sitting on my train.

By now I've started to sweat. This is one thing I didn't count on – and in Germany of all places. Things here are supposed to be orderly – *ordentlich*. Someone once suggested, rather unkindly, the difference between Britain and Germany. In Britain, unless they say you may *not* do it, you may. In Germany, unless they say you *may* do it, you may not!

We are still in the tunnel, still on rails. But now there is light up ahead. We emerge – and we are somehow back in central Essen and everything is normal. We stop and a humming sound emerges from under the carriage followed by a loud click. And all is revealed. Some clever transport engineer in Essen has introduced a bus that at certain points on its route transforms itself into a train. The bus wheels turn up and another set of wheels settle onto a railway track and it heads underground to cut out the traffic in the town centre. The speed of the tube and the flexibility of a bus. Brilliant! But they should put up warning signs for the unwary: 'This bus is amphibian.'

Why was it that the first thing I thought of was an alien abduction? There's a lot of it about at the moment – from sci-fi novels and movies and TV series to UFO-watching societies to apocalyptic cults and visionaries. And it's hard to separate out the possibility, even the likelihood, that we are not alone in the universe from a much deeper and more problematic psychological need to be a true believer. There are times when we would like to have the certainty that, despite the mess human beings make of the world, somewhere out there is a benign 'advanced civilisation' that will eventually reveal all and save us – or prove to be the source of all evil so that we can happily don our space suits and go out and smite it hip and thigh (or whatever is the appropriate term for their intergalactic anatomy). That same human need produced gods and idols in the past and little green men in the present, which raises the question about where God belongs in this mixture of experience and fantasy.

It's an old question and hard to answer. Our religious traditions tell us one 'truth', but since there are a lot of them giving contradictory messages they may not help when it comes down to the fundamental questions. Our rational minds succeed pretty well in pinning down the mysteries of the universe, explaining things – or explaining them away if we have to. Reason and science do not answer everything, but a God who is only there to 'fill the gaps' is not very satisfactory either. Emotionally we do need to respond to the great mystery of our own private existence, and the mystery of others and the greater mystery of the love that can exist between us. So what is the religious answer? I suppose it has to be a mixture of enormous faith that there is a god or that there is no god, depending on our temperament or experience, coupled with enough doubt and scepticism to stop us becoming fanatical about it. Add to that a lifelong awareness that the mystery will always remain and that we need to try to separate magic and wish-fulfilment from awe and hope.

In one view we are just tiny organisms swarming over a fragment of organic matter floating in a cosmos whose dimensions we cannot even begin to grasp. Yet each of us contains a universe of imagination and potential within us as well. As the rabbis put it, everyone should have two pockets. In one is a scrap of paper with the words 'I am but dust and ashes', and in the other the words 'for my sake the world was created'.

A final thought: try to keep your eyes open as you

travel through the world and always examine the wheels of German buses very carefully. That way, if you do get abducted by aliens, you will at least know it's the real thing.

JM

A thought

Faith is a fine invention
  For gentlemen who see;
But Microscopes are prudent
  In an emergency. (Emily Dickinson, 1830–86)

Modest doubt is called the beacon of the wise.
(William Shakespeare,
Troilus & Cressida, Act II, scene 2)

In search of a real Scot

I make it on to the Edinburgh 'Fringe' and decide not to go back to London but to have a holiday and meet real Scots instead. It's so easy to cover an awful lot of territory and never leave home. And it was like that in Edinburgh. I looked around at the lunch table after my performance. Golders Green was well represented, so was the Isle of Man, so were BBC people I knew from Manchester and the EU lot I knew from Brussels, but real Scots – 'echte Schotten' as I growl gutturally but ungrammatically to the EU man next to me – narry a one!

I remembered my hitchhiking days when I was so preoccupied with the number of lifts and youth hostels visited I never got acquainted with the lands I was passing through, rich in history and monuments though they were. Like American youngsters I'd got caught up in the numbers game.

In Edinburgh I am advised to go to a bar unknown by the glitzy-ritzy entertainment world where I am assured the real thing hangs out. There I do meet a proper Scot in a bar, which is a more serious macho place than an English pub because it doesn't have that classy southern glaze that confounds and confuses direction and purpose. I have to listen to him hard because Scots speak more basso profondo than me, more like Americans or Marlene Dietrich, so I cease chattering in Sassenach, sherry-sipping style, but gaze silently into my Guinness, earnestly imitating my Scottish neighbour.

We soon make contact and I am relieved that he seems to mean me no harm because he is brawnier than me. The fist he clenches at me is obviously political not personal. So I happily give a friendly clenched-fist salute in return because my mother's step-brother taught me to make them in Popular Front fashion in my childhood. But that was long ago and I have forgotten whether to clench my left fist or my right. I know they meant different things. Was one the 3rd International (Stalin) and the other the 4th International (Trotsky)? Or are they indications of sexual preference like earrings? Or was the right-hand one, horror of horrors, Moseley or Mussolini? Guinness and ideology confuse me and in any case I was never good at left and right – 'right' is still the hand I 'write' with.

He looks as baffled by my wavering fist as I am by his and he experimentally tries the other one just as I try to be conciliatory by clenching both fists at him. He takes a long draught and we end up waving both our paws at each other, highland dance fashion.

My English friend who accompanied me is laughing hysterically and tells me sharply to put my mits down as the whole bar is now convinced I must be a secret Irish provocateur, loyalty undetermined, and not the pious Sassenach Jew I purport to be.

To meet more Scots, 'echte Schotten', uncontaminated or diluted by the festival, I decide to give Edinburgh a miss and transfer to Glasgow, to a priory there, where I receive a warm, friendly Scottish welcome. There are lots of Scots in Glasgow and they are all friendly, as they themselves tell me, even if a little metaphysical. I ask a man the way to my priory and he informs me merrily that all roads lead to happiness, which is doubtless true but too thought-for-the-dayish to be appropriate, and that is my line not his. He then adds the material to his metaphysical directions and I thank him warmly.

Listening to the learned Scottish Dominican Fathers at their priory over breakfast I decide to reorientate myself. Instead of thinking east like any good Jew, to Minsk, Pinsk, Jerusalem, London's East End and the Polish Pripet marshes from which our Yiddishe mommes emerged, I try to think Gaelic west and bog Irish instead, into which no ancestor of mine ever sank. I even look through a Gaelic grammar and retire shattered by even more silent letters than Hebrew. Still it sounds so soft and yearning and lovely, rather like Yiddish.

The Fathers also make me realise how ignorant I am about things Scottish, despite reading history at Oxford. As with nearly all the Scots I've met, they casually refer to our English king Edward. Mentally I run through our eight Edwards. He can't be Mrs Simpson's Edward, nor the Tower of London one, nor the gay one, nor the one overwhelmed by his tougher sisters, Mary and Elizabeth. Laboriously by dogged elimination I end with Edward I who devastated Scotland and whose policy of medieval mayhem and near genocide was summed up and dismissed in less than half a paragraph of my school textbook.

One of the Fathers is preparing a book of Scottish poetry before 1350. I didn't know there was any. But I look at his proofs and there was obviously a lot and very fine stuff too, moving and witty even in translation from Gaelic, Norse, Welsh, Angle-ish, and probably Pictish if they knew how to read it. 'Scottish Welsh?' I enquire bemused. He nods over his Bran Flakes.

I don't argue because I have discovered to my shame that I've lost a whole kingdom, which lasted three centuries, and ruled the Isle of Man and the Western Isles. I had never heard about it till now and yet it is or was another kingdom in the history of the United Kingdom of which I am a citizen. Why were we never taught 'United Kingdom studies' at school? Why were we brought up to be so ignorant of our nearest relations and neighbours? Why have we never read their epics even in translation? Why have we only mugged-up the highland clearances and the Irish potato famine for exams but never learnt to feel them?

I realise with shame I know nothing of the spirituality north of the border and that a quick run round of the Isle of Iona by coach will not reveal it. It will be like the modern 'medieval' banquets with finger-lickin' chicken and pizza that serve as culture on the Costas. And spirituality requires even more effort than culture. No one else can pray your way to it other than you. Hoots!

As a spiritual exercise, have you ever dropped into a local chapel on holiday and meditated there more than five minutes?

Have you tried to see the Holy Spirit as well as a museum brochure at the collections of religious art? Have you tried to open up to that art and find out what it has to say to you?

Since God is everywhere, where is He in your holiday resort? It's worth finding out. You may need Him.

Have you met locals as well as tourists? Where do they go for their holiday? To Britain? What do they find here?

LB

A joke

I ask if there are any Scottish religious jokes. After intense thought one is produced. A man deep down in hell cries up to heaven above, 'Lord, Lord, I dinna ken, I dinna ken! The Lord peers down and answers tartly, 'Well ye ken the nu! Well ye ken the nu!'

I shiver!

Street opera

There's something missing from our urban streets, and you have to go elsewhere to find it. An old French musical film begins with the noises of a street at dawn, the sounds gradually coming together to make a kind of symphony. It must have been a bit like that here in past centuries when the street-sellers used to sing about their wares – an idea taken up in Lionel Bart's *Oliver*. But today, even newspaper vendors have stopped shouting out their incomprehensible 'Read all abaht it!' With only one evening newspaper there's no one to compete with for sales.

But I know two places where it is still possible to hear this kind of street opera – natural, unorchestrated and quite beautiful.

The first I heard on a train going into Madras. It was a last-minute invitation to a conference in India. Just enough time to get the shots and papers and catch a plane. I don't remember much about the conference – except for two humbling experiences: the young man who wanted to hear my lecture on Judaism because he was interested in the 'minor religions'! And someone else who explained that from the perspective of the Indian subcontinent, Great Britain was just a small island off the

European peninsula of the Euro-Asian land mass.

But on that crowded train, with hard wooden seats and people hanging off the sides, I heard a series of overlapping women's voices as they passed through the compartments selling their wares. Each sang her own melody, to unknown words, but the total effect was entrancing as the waves of sound grew louder and passed to be replaced by the next. I could not even begin to imagine the poverty and suffering that lay behind the lives of those singers, or with what weariness they trudged through those crowded, swaying carriages – but the beauty was breathtaking and transforming.

The other place was Jerusalem, and a most unlikely musical treat. The *yeshiva* is an academy where Orthodox Jews spend their day studying the huge volumes of the Talmud. They use a unique method of learning, working in pairs, arguing fiercely over the meaning of each and every word. So, most of the time you can hear only the rumbling and sometimes roaring of masses of argumentative male voices. On the inside it is a cacophany.

But once I stood outside a *yeshiva*, and the background rumble was muted, and individual sounds came through. Many were chanting the passage before them, a kind of sing-song intonation that is also a way of interpreting the meaning. As they argued, one voice would rise above another, or from another corner someone would let loose with a beautiful tenor chant, only to be replaced instantly with some deeper voice in disagreement. Again it did not matter what was being said – it was simply an operatic sound, something utterly beautiful that would embarrass those taking part if they ever learnt what an effect they were having.

The Psalmist wrote that the trouble with idols is that

'they have ears but hear not', and those who worship them become like them. So we have a *spiritual* task to keep our ears open as well as our eyes, and to be alert to the melodies all around us. Think of the ones you know from your own place or have discovered on your travels. Some are the voices of people at work, others, like the call of the muezzin, are of people at prayer, but all that move us speak to the soul within us. The human voice is an exquisite instrument, especially at those moments when the one who sings is utterly unselfconscious about the power of what he or she is doing.

Since we abolish buskers in our public spaces, and would probably put a ban on any *yeshiva* in England that made such a noise, we may have to travel abroad to hear such things. The best we can offer here is competing ghetto-blasters, or the thumping bass coming out of someone's walkman. But maybe I haven't learnt how to listen to the sounds of London. Maybe our streets are also alive with the sound of music.

LB

A thought

> Open the eye of your heart, so that you may appreciate what you see with the eyes of your head. (The Foundation of Religious Awe)

Your finest prayer

Some sages said that the finest prayer is a song without words. You just sing 'la la la' or 'li li li' to the tune in you that wells up from your heart. That way you can pray in your bath or under a shower in your en suite bedroom.

Lost music

A century ago the streets of London and other cities were alive with the cries of hawkers and itinerant vendors of goods and services. The stereotyped repetitive cries were chanted musically in many instances. Typical cries were: 'Hot pies! Hot pies!' 'Buy a broom!' 'One a penny two a penny Hot Cross Buns!' 'Any old chairs to mend?' 'Small coals here!' 'Water cresses, fresh and fine!' 'Any knives or scissors to grind?' 'Sweep ho!'

These have now all died out except perhaps in some exotic foreign places. The last to go were the familiar cries of newspaper boys selling '*Evening News*' or '*Evening Standard*', sometimes with a shouted sensational headline.

In my own lifetime I can recall mainly variant cries of the rag-pickers and lumber men: 'Any old rags or lumber?!' The place of street cries has now been taken over by expensive television commercials, with irritatingly repetitive musical jingles, hawking everything from motor cars and chocolates to cosmetics and sanitary towels. Perhaps a near relation of street cries is the surviving call on the London Underground trains 'Stand clear of the doors!' but now delivered by a cultured female voice on a recording over the carriage speaker instead of a raucous chant. (Leslie Shepard)

Strange holiday – the real thing!

The loveliest holiday I ever had was in hospital and not an expensive private one but an ordinary NHS one. I don't know how I got into it, though when I came to the nurses told me I had had a fit outside a joke shop – a full-scale epileptic one, grand mal. I had gone rigid and purple like an aubergine and then crashed my head against the kerb stone. This may sound horrific but fits are a spectator sport and I was unconscious of it until I came to and the doctors were doing tapestry work on my skull.

They advised me to stay in over the bank holiday and I wasn't going to protest because the rest of my life was not going that well (hence the fit) and I badly needed a holiday from responsibility and from myself.

Most of the doctors were away over that weekend and there were few admissions, so an atmosphere of peace descended on the wards. I had no visitors, my friends were either away or had not heard, and I had no desire to tell anyone where I was or what had happened and cope with their concern. I was content in the company of the other patients in my half-empty ward. I enjoyed the anonymity of worn regulation nighties and the absence of fuss.

We lay torpid like the bodies on a Costa, occasionally offering each other sweets or advance warning of the tea trolley. Some were quite ill, far worse than me, and I was touched how nice they were to me and each other. I had expected their problems would make them tetchy and difficult but it was the opposite. Their problems had brought out their original goodness and I found myself blessing God for the care in them. We all knew the score about each other and were tactful about it.

One instance. I had come in without the normal toiletries, books or bags. They shared their Indian savouries with me and gave me squirts of aftershave and even offered to let me borrow their glasses because mine had got smashed when I fell.

The days of that weekend passed like a dream, reading the magazines you find in dentists' waiting-rooms, gazing at London town distanced by the thick glass of the windows, with all its restlessness and passion softened.

At first I mused about the might-have-beens of my life and then fantasised about what would have happened if I had chosen differently. These fantasies then came together in stories, and effortlessly the plan of my first novel (as yet unfinished) formed in my mind.

I realised I was not particularly attached to this life. It had its moments but I had experienced too much depression and anxiety with them to want it to last for ever. At some time I would be willing to call it a day because I had gone through it. I had done it and was curious as to what lay on the other side, if it had another side.

I managed to wander down to the hospital chapel where I just sat and invoked Whomever Whatever and waited patiently for whatever happened or didn't. It had a good atmosphere because it was continually sanctified by the prayers of those who meant business and weren't just repeating words. So its quietness was no emptiness.

Like many patients I thought the nurses were angels. I certainly didn't encounter a grumpy or vindictive one. When the doctors returned and the bank holiday was over, I realised the weekend had given me all that I had ever wanted from a holiday, without frenetics or over-strain. I said this to a nurse who at first said she was astonished but then admitted that the calmness of such weekends touched her also.

After my partner and I broke up, I went on a commercial club holiday to recover. The sun and the pool were satisfactory. It cost a lot and was not nearly as satisfying as the hospital. There I had moved a step nearer eternity and I liked it. It was an adventure.

Thinking spiritually about holidays I realised how often we recreate the same tensions, the same situations, the same dramas that we are trying to escape from back home. It is the same with synagogues and churches. Lots of tired businessmen join them as an antidote to the competition and combative materialism of their working lives. But as soon as they get to church or synagogue they are unable to break the pattern, the compulsion, and throw themselves into fundraising drives, constitutional subcommittees and competition for office. They recreate the very world they were trying to escape. And it is the same with ordinary holidays too.

During my hospital 'holiday' I found the answer to problems, not just their repetition. I tried telling people this but they didn't believe me and thought it was just the tail end of my fit. But thank you NHS, you touched me with the truth that had almost passed my understanding. I had needed my fit to know it.

It is not easy to give oneself a real holiday. You only repeat your life in a different latitude. What is a real holiday for you – the real thing?

LB

A question for you

Try to think of the times when you really found peace – not when you ought to have found peace or other people told you this was peace. I do not think I am the only one who has found it in acceptance and illness.

LB

My soul chooses a holiday resort – Liverpool

So they're trying to refashion Liverpool, that self-destructive old whore of a town, as a holiday resort with trendy, empty modern art and sentimental icons of the Beatles. Good luck to them! It will add yet another layer of paint to its raddled Georgian-cum-concrete features. Why not tidy up Toxteth genteel-like – that would be a laugh.

I sit in Lime Street Station wondering why I've come, though I'm pleased to be back. And I hum the dirty ditty of a none-too-clean town, consuming fast food and chips.

> O dirty Maggie May, they've taken her away
> and she won't be walking Lime Street any more.

Nor will I sadly because I'm not permitted to drive up to Liverpool any more (epilepsy) and the little house I used to stay in, like me, is not getting any younger and is now a memory.

I loved Liverpool because it never put on the style like London. I liked watching chocolate bars dipped in batter, fried in the fish and chip shop, and mothers force-feeding their kids with pies and pasties, and the little old man slipping a half a lamb into his grubby raincoat and scuttling away before I cottoned on. I've never seen an OAP nick half an animal in London.

A hard life had given the people a good heart. I believed the check-out girls who wished me a happy day. I think they really meant it and they were wonderful fishing me out from the freezer when I fitted, epileptically-speaking, into it. I also learnt in bars and clubs how to call out Kelly's Eye and Legs Eleven and Unlucky for Some and Two Fat Ladies. Not many rabbis know that.

Not putting on the style was an antidote to religious role-playing. Like the taxi driver who looking up saw the high clerics on top of the Roman Catholic cathedral, Paddy's wigwam, glumly looking over the parapet for repairs. 'Don't jump, your reverences!' he shouted tearfully. 'It can't be as bad as that!' Or the curate in the Anglican cathedral who saw a vision of God Almighty in the pulpit. 'What shall I do?' he cried to a canon wringing his hands. 'Look busy!' replied the canon thoughtfully.

When I left Liverpool last, a lot of red and white English flags were flying in Liverpool, for it's a sentimental town and a football one. But now that Northern Ireland, Scotland and Wales are beginning to go their different ways, it's worthwhile working out what all those St George's crosses mean to us besides football, because I think we'll be waving a lot more of them in future as England seeks its own identity.

For me those flags stand for the very English, very Victorian, unofficial, female, gutsy saints I discovered on Merseyside. They live in a barred side window in the Lady Chapel of the Anglican cathedral. In a time when there are so many kitschy saints, pop stars and princesses and popes, they are the real thing, women who unsentimentally gave till it hurt. You don't find many of them around these days. There is the one who in a cholera epidemic washed the infected clothes of the sick in her boiler, and the one who without fuss went down with a ship when the lifeboats were too full, and my favourite, Josephine Butler, who fought for those Maggie Mays in the song all her life, working to save them from brutal arrest and horrifying medical examination, while nobody bothered about the men who used them. She used to take the diseased, battered wrecks into her fashionable home and nurse them with her husband and servants in the best guest bedroom till they died. That's the real English religion that cross stands for.

Occasionally my soul overrides my body and my mind and says it must have a holiday too. Its demand is so strong the other two are bewildered but forced to give way. My soul doesn't want culture or history or religious respectability or professional piety or conducted retreats or sun oil or beautiful people, just an earthly reminder of what religion is really about, which is easily lost.

My uncanonised lady saints were very Liverpool and very tough. As tough in their way as the legendary Dixie Dean. 'Who's Dixie Dean?' you ask. 'Man, where have you been!'

They're the reason I've come back.

LB

Spiritual exercise

Which are the unofficial holy places of your life? Do you ever revisit them?

A thought

Even some old-time rabbis were willing to hang out with less than holy people:

> Some lawless men lived in the neighbourhood of Rabbi Ze'era, but he was friendly towards them in order to win them over to repentance. The rabbis, however, did not approve of his conduct. When Rabbi Ze'era died the men said: Up till now we had him to intercede for God's mercy for us, but who will do so now? And they felt remorse in their hearts and repented. (Talmud Sanhedrin 37a)

JM

The other Majorca

I had been packaged to Majorca twice before I discovered the Other Majorca, which does not mean more sex and booze – less of both in fact.

The key to alternative Majorca had been staring me in the face, but as so often in life we do not see what is in front of our eyes. In this case the key was so obvious, I scratched my head wondering how I had missed it. It is the great statue dominating the harbour of Palma of a near saint who never quite made it – Ramon Lull.

You've never heard of him? Never mind. I hadn't either, but once he was one of the most famous intellectuals of Europe. Entire faculties of universities were dedicated to Lullian studies.

Sometimes if you wander round an old priory library, you may come across shelf after shelf of his collected works – all in Latin with crumbling, worm-eaten covers. He was the rival of the great Thomas Aquinas himself and his aim was not so different – to synthesise all human (and divine) knowledge into one great awesome system. A bit like Marx, I thought, centuries later.

But being a modern with shaky Latin I became more interested in the man than his theories. One day when I am older and greyer I might have a go at his theories because I'm curious – but later, later...

He fascinated me. According to the stories I heard in Majorca told me by monks in almost empty monasteries he was once a chic, trendy, late-thirteenth-century, womanising squire, who thought nothing of a bit of adultery if he could get it, like many of Majorca's visitors today. He chased one married lady so relentlessly, she finally invited him to her room. Opening her blouse she exposed the breasts he had so long lusted after. One was already eaten away with cancer!

The story goes on that this shock sent the young squire into spiritual free-fall as it would any of us. He forsook this world for God, the Church, religious study and contemplation. The passion in his nature never left him. Years later it blazed out again cruelly, when in a fit of anger he struck and killed his faithful Muslim slave, who had taught him Arabic and introduced him to the Islamic civilisation on the other side of the Mediterranean.

This passion could also turn from cruel to heroic because he also sold himself into slavery in Tunis when he was an old man, to redeem another Christian prisoner. He died a martyr. He was a very great, passionate and uneven man, and I would have liked to have known him, though a bit wary.

I became quite interested in him and, fed up with bars, sing-songs and sangria, tried to track him down. It wasn't difficult. His spirit inhabited every high hill and peak on the island. A network of Lullian monasteries and hermitages looked down on the holiday hotels. Some were thriving resorts themselves, like Blackpools of the spirit. Coach-loads arrived for a quick religious fix, a scamper round the rosary, a reduced-price souvenir, a blessing, and a honey nut cake. Some just sold postcards and ice creams. Others were deserted except for an old friar or two. You could lay down your sleeping bag and put whatever you wanted or nothing in a box. Others were just deserted. I found one of these in a field hidden by high ripening corn. It was the most peaceful place imaginable, a one-storey building encircled by a veranda. I sat on and on until the sun set making friends with the gentle ghosts or spirits of the place.

I visited it with a matter-of-fact friend. He had been an undertaker and had little time for gentle spirits, but he too remarked on its peace. Even after I arrived back from my Costa the memory of it still lingered. It lingers in fact till this day and has never left me.

The soul is part of us and leaves its mark anywhere and everywhere. It will give itself to you, if you give a little time to it. If you peel back the material world, a spiritual one is revealed. Just a little perseverance!

LB

People

The patroness of package holidays

Now Margery Kempe ('poor Margery Kempe' people say superciliously) was not the greatest second-class saint I ever bumped into while reading English history but she was certainly one of the oddest. And at the time I bumped into her we had a lot in common. She was neurotic and so was I. She liked men and so did I. She'd caught the travel bug and so had I. She must have been bored stiff by the Wars of the Roses – because though she lived through them she never mentions them – and so was I.

But don't get me wrong. I'm pleased five centuries separate us because I couldn't have stood her for a day let alone on a long voyage. For one thing she cried non-stop. Whenever she thought of Jesus, which was often, the floodgates opened. And this was particularly distressing at meal times. Medieval pilgrimages were the precursors of modern packaged holidays, and her party couldn't stand it. Rather than face another breakfast with her, they stole away from the inn before dawn without her. Dumping her was not nice, though understandable, because it was risky being a lady traveller alone, then and now.

However, they had their come-uppance at Venice. Seeing her on their boat to the Holy Land, they promptly changed boats – and felt very pleased with themselves till she said she wouldn't go on their boat for all the world because God told her it was going to sink. Whereupon they all fell over themselves trying to change back again.

But there must have been something about her because just when she was abandoned somebody always came to her rescue. There was a kind Devonshire man called William who saved her in Central Europe and an Italian Princess who befriended her. Her autobiography only survived by chance. It was lost for centuries – the hand-writing was so awful nobody could be bothered to decipher it until 1936 when it came to light in a library and was recognised as 'The Book of Margery Kempe', which people knew about but no one had read. It now lies in state in the British Museum.

I like her too because she was so honest. After, not before, her religious experience, she dated a married man behind a church pillar in King's Lynn. But he was only playing with her and told her she was just filth. Now I know many saints who have admitted their sins before their conversion, but no one who confessed to being stood up after. I respect her for telling it straight without excuses. She even told God the Father, who wanted to marry her mystically, that she would rather marry Jesus who was younger and used to sit at the foot of her bed in a purple robe.

She also mentioned meeting the great mystic Julian of Norwich. Margery had a broodful of children whom she never mentions at all and Julian, who probably never had any, talks about God our Mother. I'd have liked to listen in on their conversation. Margery, I guess, did most of the talking. She was a very courageous talker, even daring to tell the Archbishop of York, who was not one of her fans, that 'he was a very naughty Archbishop'. She was lucky she escaped alive.

As well as crying and talking, Margery got around. She was inspired to visit Jerusalem, Rome, Canterbury, Santiago de Compestella, Danzig and Eastern Germany. Why Eastern Germany? I don't think she knew the answer to that either.

In Rome, on her way back to England, she came across a poor woman with sick, infested, starving children. Many modern tourists encounter destitution on their travels, but few of us act like Margery. She delayed her departure and stayed with that woman for months, cleaning out her hovel, selflessly disposing of the filth and looking after her and her children. She never regarded such care as the purpose of her journeys but I think God and you and I see it differently. It was not the holy places but the holy place inside Margery that mattered. Perhaps that was why she was inspired to visit Rome in the first place.

When you get back home from your holiday, try to see it as God sees it. You probably met God there, disguised as a beggar or hotel bore or somebody you gave your seat to in the overcrowded airport, whose name you no longer remember if you ever knew it in the first place. You might be startled by such a different point of view. From your point of view, of course, it was a holiday, but from His point of view it was also a Holy Day, when heaven came close and unexpected.

LB

Holiday romance

My friend returns with our breakfast papers. I wrap my broadsheet round me, while he studies football in his tabloid. But while he's looking for the marmalade, I sneak away his tabloid to read the agony aunt's advice, which is sensible and gripping. As the agony is love, the crisis often erupts in bed, the man bare to his navel, and his love – or is she? – in her best bra.

Now, if we all enjoyed loving committed relationships, we wouldn't need agony aunts. But that's perfection and millions have to spend all or large parts of their lives without them.

Suddenly something clunks through the letter box, and whilst I retrieve it my friend reclaims his paper. I don't mind because a packet of those luscious, illustrated sunny summer holiday brochures has arrived which millions of us peruse as we plan our annual trek south. The holidaymakers are so loving, so committed, as they gaze at each other in their pictures.

But I'm not convinced. People on holiday gaze at each other for many reasons. They want to make love because the needs of their bodies are overwhelming, because they're lonely and it's a way to make contact, because they get warm that way if their Costa is cold, because sex works like a sleeping pill or tranquillizer, because they're scared of missing out, because it's fun, an escape or an addiction.

Can religion give as practical advice as an agony aunt? Well, religion means responsibility, so make sure you know about safe sex and have everything handy if there's any possibility of it happening. (The Talmud warns us not to trust ourselves till the day of our death.) The safest sex is, of course, no sex, but be realistic. Religion also means kindness, which doesn't mean fair words but keeping promises.

And remember, light holiday relationships easily turn into serious ones. If you want deep satisfaction, feeling somebody isn't enough, you'll want to feel with them and then for them. And then you're beyond sex and into love. In which case ponder these two quotations from a prayer-book: 'Love is a decision not a feeling'; 'You earn love, as you earn your living, by the sweat of your brow.'

A lot of sex is in the mind not your genitals, especially holiday sex. How often does one hear a variant of this conversation overheard on the plane home? 'I felt like crying. I can't bear to be without him. I know what – I'll write to him. Now what was his name. Surely you remember. It ended in an O – Pedro or Rinaldo or Antonio or something. O dear!'

All this painful wisdom is distilled from years of running a religious divorce court, from hospital visiting and my own muddled loves. I don't need it personally now, having achieved a loving committed relationship in my latter years (which doesn't mean no rows), and in any case I'm beyond it. As one old Jewish woman says to another: 'So your husband's gone to Hawaii on business. Aren't you worried about those young chicks he'll meet there?' 'Look', says the other placidly, 'at that old dog running after that car and barking. Do you think he knows what to do with it if he ever catches up with it!' Well I'm that old dog and I was a religious agony uncle once on TV myself.

A cautionary tale. If holiday sex is a must, then you have to think carefully about where your lusts are likeliest to be satisfied. A girl I know, a remarkably good-looking girl, booked a last-minute holiday to a Greek island. On the first morning she strolled along the beach where handsome muscular men played volleyball. They were polite to her, too polite she thought, and courteous, but something was missing. She retired to a beach café puzzled. Listening in to the conversations around her, the penny dropped. She was on holiday in the gay capital of the Med. No wonder her high-cut bikini didn't even rate a whistle.

When she got back she told me she had quite enjoyed her time among the gay men. In a kind of way it was a relief not having to have IT. Also she was much in demand as 'auntie' and was invited to all the parties – she was so safe. She now knew she said what it felt like to be a gay man in a heterosexual world. She wouldn't have missed it for anything but she wouldn't go there again for a few years or until she had a very steady, straight partner.

LB

A question

What's the difference between love and being in love?

Alone

Conversation is often a means of preventing contact between people, the opposite of communication. Running away from silence is a way to avoid real listening to what the other person is feeling and contact between you deeper than words. The continued presence of other people around you is a way of avoiding any meeting with them or your true self. Just as babbling too many prayers is a way of preventing God from getting a word in edgeways.

That is why we frantically find 'friends' on holiday and throw ourselves into any activity going – why we are scared of just letting aloneness happen and tasting it. (I say 'we' because it used to terrify me.) Why is it so frightening to us? Why not taste it? Aloneness is not loneliness and can be the biggest adventure of all if we can cope with it.

Aloneness can usually be ended with courage. A friend of mine found himself dining alone at a hotel abroad. He approached another man also dining alone. 'Since we are both dining alone, I wonder if I could share a table with you. It might be agreeable to us both?' 'I have no wish to share with you,' said the other politely. 'I am content as I am.'

I asked my friend what he did then? 'Oh, I went up to another chap and said the same thing.' And what happened then, I asked admiringly, because I could never have coped with one such rejection. 'He said "Delighted dear man. I hate having to read the paper at dinner every day." '

If you have enough bravura most times you never need be alone. But how many of us have it? I certainly don't and it's easier for a man than for a woman still.

If you do not want to holiday alone, then plan early. Do not leave it to the last moment when you will be surrounded by couples playing happy families and charged exorbitantly for the privilege of a cubbyhole by the lift shaft.

Consider the following suggestions.

If you do not want to spend the Christmas holiday alone, book into a retreat house which is having a Christmas house party. They are very quickly booked up so act fast. You will usually get a small room, cheery with a washbasin. You usually won't get it en suite – but then you can't have it all. There are facilities to make tea, and chat, friendliness and company, good meals (luxurious school dinners type) and often a bar.

Some of the big commercial companies have single weeks with no single supplement, but not at most popular times in high season. There are usually more women than men, but the former of a higher calibre than the latter. Ask if they do line dancing, then you don't have the torture of being the only one left out on the dance floor.

Special-interest holidays are a way of meeting other singles gracefully because quite a lot go on them, as well as couples, and even the couples do not radiate 'keep-out' warnings. There's rambling, painting, holistic therapies, and someone is always needed to supply a fourth for bridge, and the more your new partner loathes you, the more your opponents will like you.

Pilgrimages can be very matey – and not only with God.

There are also holiday company house parties especially for solos. The companies do try to balance the sexes but it isn't easy. Don't fib about your age when applying. It leads to complications. Confide in the company if you are an exceptionally young 'with it' 60, 70, 80 or 90. They've heard it all before. You're not so singular.

But if after all this you do end up alone among happy families and couples, and don't have that magic bravura or 'chutzpah', here are more suggestions.

Be a good listener! Everybody wants to talk, few are able to listen. Show sympathy and wrap up your criticisms, at least initially, in cotton wool. Listen with a good heart. We all need someone who is on our side.

Don't suffer from false pride! If you want to meet people, tell the package holiday rep. Unless you're very unlucky and it's their off day, they'll respect you for being direct and organise something.

Be helpful! The world has far more takers than givers, so you will have scarcity value.

People-watch and let it be known you are writing a novel. Most will want to find out if they are in it. It is compulsive. When they ask you ever so casually, look non-committal but kind withal.

Sketch or paint, even if you can't! Try abstact expressionism if you can't draw for toffee.

If none of these work, then there's always God. A conversation with Him is probably long overdue and now may be the time to start. Perhaps He arranged it so. Now there's a thought! If He exists and loves you as the religions say, why not? In the end we die alone. '*On mourra seul.*' This will seem less terrible if we anticipate it.

Also, it is good if you don't let other people have too much power over you. If you believe in yourself other people will believe in you.

LB

Practical thoughts

Remember that you can be even more alone in a crowd. Being attached to the wrong person can be real hell not just limbo. Work out honestly why you are alone if you dislike it so much. Are you expecting the impossible or being too choosy?

LB

Anthology

Here is an extract from a letter written to me by a friend, a widow, who goes solo on coach trips.

> It takes courage to venture away from home at all on your own, especially abroad. But I have always found someone willing to make room for me at their table for the all-important evening meal; preferably a couple or two ladies travelling together as friends – a third is often welcomed as an enlarging of the conversation potential. Joining another single can present a 'clinging' problem and I have become a

little wary of being eagerly pounced on with the words 'Are you on your own, too? We can team up together.'

A river cruise is ideal, where largish tables are the order of the day. So you immediately become part of a small group. After this, the main hurdle is behind you. The rest is usually child's play.

I favour a moving holiday, either river cruise or coach. Best of all, a Pilgrimage group or a shared interest party of some sort. That way there are plenty of new things to do and see and discuss at dinner and I find the 'free time' most enjoyable. Usually just sufficient to explore lovely views, beautiful buildings, interesting shops or quaint streets. A bonus I find to be totally free to go whichever way I please and thoroughly spoil myself. No second person to pull the other way.

Lunching alone, especially abroad, can be daunting – but there again, usually one meets up with some of the others in the party.

I have found it more difficult when taking a static holiday. In a 'posh' hotel, wine waiters often turn a blind eye to ladies alone and one, of course, is firmly ushered to a table for one. For some strange reason I felt embarrassed, almost guilty, at sitting in this lone state. If coffee is served in the lounge afterwards, the situation improves as people gather round and chat.

A coach package holiday solves the eating alone problem but it can also land you with unwelcome companionship, so I have learnt to weigh up the possibilities pretty smartly and make a quick move to join the group that looks the most promising.

Sun, Sand and Soul

Having a favourite local hotel where you know the staff are friendly and fellow guests usually pleasant is useful. A weekend break at such a hotel can boost your morale without causing too much stress and hassle. I have such a hotel in my sea-side town – a little old-fashioned but with a feel of past grandeur and I have found much good company and pleasant conversation in this old favourite.

I have often felt jealous of the husbands and wives sitting companionably together over dinner, imagining inspiring and witty discourse, but if I stretch my ears and listen in, I find the conversation, in fact, often so mundane that I lose my envy and relax back and enjoy the food and the view.

The problem of not having someone to mull things over with or grumble at, is insoluble and I miss my entertaining, charming husband more and more as the years go by. But my choice is to stay at home and settle into grey depression or take my courage in my hands and venture forth to places I've never been to and count the blessings offered by such a variety of tour companies and vast assortment of holidays available. I can't wait to pick up next season's brochures.

LB

Staying alone

There's a narrow line between being alone and loneliness. And all the difference in the world between taking time out to be alone, away from the place where we are at home with loved ones, friends and acquaintances knowing that we can return to them, and a chronic state of isolation. Even knowing that 'in the end we are all alone' may only take the edge off what for some people is an ongoing painful reality.

But once in a while it may not do any harm to experience what it feels like to be by ourselves, away from the things we take for granted within our familiar four walls. A holiday alone can be a chance to step back from a busy life, take stock and explore our own inner resources. If we are gregarious and find it easy to meet people, or simply prepared to barge in and take a chance on being rejected, then we will find ways of meeting people in the most unlikely places. But if that is not how we operate and we're never sure whether we can take that extra step, or really want to, then we have to resign ourselves to a different kind of situation, but one with its own compensations.

I have always taken holidays by myself as well as family ones. People think it odd, but I've enjoyed the chance to have my own space for a while without the kind of daily commitments to others that family and work demand. It's selfish, but it has a price in loneliness or sheer boredom. Still, if you're going to be lonely and bored it is nice to be somewhere where it is warm and there is sea and scenery to distract you and people passing by to entertain you.

Most of my holidays are cheap, out-of-season breaks when the resort is pretty empty anyway and not much is going on. I do a lot of walking at first to explore the area. It makes sense, of course, to see what is there, but I think that I am also marking out the boundaries of this new territory rather like animals do. It is obviously some deep need to take control of my environment and create a kind of artificial familiarity and security.

I make sure I am well stocked up on detective novels and usually remember to pack a radio. The BBC World Service is always a comfort, but wrestling with static, fading stations and obscure languages can be wearing. I have a couple of fall-back options for entertainment if these begin to wear thin. One is my notebook – to work on some unfinished project or poem. It's good to have to use pen and paper even though I'm hooked on my word processor. Some people take a sketch pad for the same purpose.

And then at the bottom of my suitcase are a couple of harmonicas, the ideal portable travelling companions. If I'm blue, out comes the blues harp and I can blow it all away, sitting on the side of some isolated sea-shore. (If you're going to feel miserable, really indulge it!) And if I'm just lonely, nostalgic or sentimental, out comes the chromatic (even though I still can't use the slider that changes key properly) and I can work my way through forties love songs – provided there aren't too many sharps or flats.

An instrument of any sort, as long as you can transport it or find one locally, is also a great calling card for meeting people wherever live music is being played. I've found that I only need one party piece on the harmonica to get in the door, then I can sit it out on the sidelines and enjoy the rest of the music. (In Munich years ago there was a 'folk song club' and anyone who could do a chorus of 'Ilkley Moor Bar t'At' got free admission and as much beer as they could drink. Suddenly Munich was filled with authentic British folk singers.)

Buried in a good detective story or exploring new territory or checking out different restaurants I can get lost in another world. But then comes the moment when such activities or distractions don't help and I'm suddenly faced with the raw fact of being alone. The more my mood veers towards misery and self-pity, the more ironic I find myself becoming. This is what I asked for by going away by myself, so what am I complaining about? At this point another mechanism cuts in. I might find myself humming the tune of a song by Bertholt Brecht, roughly translated: 'As you make your bed you must lie there!' And that brings a smile and enough detachment to take the edge off the mood.

Sometimes I'll find myself becoming quite spiritual, counting my blessings and pointing out to myself how lucky I am to have such an opportunity to get away. At others a brisk walk can take me out of my mood. And sometimes I simply have to stick with it and see where it takes me or what it is telling me about myself. Stepping out of my daily life and the hustle and the politics and the busy-ness does give a chance to reflect on my own journey so far and where the next steps might take me. It's an awfully elaborate mechanism – packing, planes, resorts, hotels, etc. – just to get a few real moments of introspection, but it's cheaper than a therapist and more effective than some of the services I attend which are supposed to offer the same opportunity to reflect.

Being alone, especially in the evening, I feel particularly conspicuous and self-conscious, strolling along the front or past crowded restaurants. Everyone else is either paired off or with a family group. They always seem to be having a good time, even if they're not. And not having a good time in a crowd still seems better than the same experience alone. Mostly I ignore the feeling but I'm saving one solution for when I'm really desperate. Since it is more a problem of imagining what other people think, I can always take up jogging. A person who is obviously all alone might evoke pity. But a person jogging is obviously an athlete and worthy of respect!

If I've been alone for some time with no one to talk to I think of it as a kind of language 'diet' or 'fast'. After all I talk enough the rest of my life and most of it is pretty trivial stuff, so why not shed a few verbal calories once in a while? Maybe I'll appreciate words more and use them more carefully when I get home. It can even turn into a kind of dare to see how long I can go on without talking at all. Which is perverse, but can also be quite fun.

The danger of being alone too long and with too much introspection is that you might end up feeling you've disappeared. That's when it helps to think about Yankel. Yankel was so absent-minded he could never find his clothes in the morning, so he hit upon a great idea. He wrote down on a piece of paper exactly where everything was as he took it off at night, and then he tied the paper to his wrist. The next morning he looked at the paper and there it all was. 'Yankel's jacket is in the cupboard.' He looked, and there it was. 'Yankel's trousers and shirt are on the dresser.' And so they were. 'Yankel's underwear is on the chair.' Yes! The last note said: 'Yankel is in the bed!' But when he looked, Yankel wasn't there!

JM

Staying with friends

Staying with friends makes me nervous. Perhaps at the back of my mind is an old proverb to the effect that 'visitors and fish don't keep' – or was it that 'after three days both start to stink'? Sorry to be so crude about it, but I always feel that I am intruding, and become conscious of negotiating my way around a whole series of embarrassments or potential traps.

For example, why is it that whenever you stay with a friend, or worse still, the friend of a friend who generously agrees to put you up for a few nights, the toilet paper runs out while you're in the loo? You know from looking at it that the roll must be near its end, but you figure it will just last out this occasion. It never does. Apart from the immediate problem, you then have to ask for more, which seems to be a kind of reminder to them of their poor housekeeping – or your inordinately greedy consumption of rainforests' worth of paper. Either way it's a loser. Of course they do keep a supply in the bathroom somewhere, behind one of the cupboard doors or secreted behind the plumbing – but are you allowed to poke around looking for it? Is this an invasion of privacy that is going too far? Give me a hotel any day.

A footnote on the above. I had a breakfast invitation at

a prestigious hotel in America. It certainly has a lot of marble. American 'washrooms' are usually well looked after, though the western-saloon style loo doors are a bit too high off the ground for my taste. (Which is probably why so many people seem to get shot in them in American movies.) But in this marbled palace with ceiling-high mirrors and chandeliers my cubicle was paperless. Fortunately my innate distrust of such situations made me check and peek next door before starting. So I grabbed the half-used roll from there. Perhaps it was too early in the morning for them to have tidied up, or maybe there was a rash of loo paper thefts in New York. Nevertheless it bears out something a businessman once told me. When visiting a company you wish to do business with, the first thing to do is check out the loo. If that is well looked after you can trust the rest of the organisation.

Meanwhile back at my topic, another hazard of staying with friends is the blocked bathroom sink. Presumably months of hairs and other unmentionable materials have 99 per cent filled it, just awaiting your visit. So should you try to clear it – wading through decaying remnants of other people's goo? Or do you give up in defeat and simply sneak out, hoping the water will seep away or that someone else will go in there and unplug it without comment? What etiquette book gives guidance for events like this?

Almost as difficult at times is learning how to navigate the hazards of someone else's furnishings. Low-lying shelves ('we've been meaning to take it down for ages') take half your scalp off when you enter the spare bedroom; bedroom lamps with funny switches fuse or fall over in the dark; sharp-edged bits of pretty furniture snag clothes. These things we all take for granted in our own home and automatically negotiate around without a second thought, are guaranteed to attack others. How often can you say: 'Oh, it doesn't matter!' 'I have another pair of trousers!' 'It's only a little blood!'? Of course such minor matters can help cement a friendship – but there has to come a moment when yet another apology for a bit of broken china or the picture that fell down in the night becomes too embarrassing.

If inanimate objects are a potential problem, then moving ones increase the chance for disaster incrementally: the pet dog, a St Bernard at least, that falls in lust with you ('you don't mind dogs do you?') or the cat that defies you to sit on her favourite seat.

And then the children. Children, other people's children, present a double challenge. After all it is their house so there is no need for them to respect your privacy in the same way as their parents would. There is nothing wrong with that. In fact, playing with children can be great fun – especially when you have no responsibility for them during the rest of the day. But their staying power when they're locked into a game is infinitely greater than yours. So what is fun for you for twenty minutes, as the same trick is repeated, or the same song chanted a hundred times, can turn into a kind of Chinese torture.

I am sure that sounds worse than it should. A lot depends on your own state of mind and what 'holiday' means for you. Of course, if you're sharing the burden of your own children with friends in a similar situation it is totally different. Then you're simply doing a trade-off and tomorrow will be your day off and their turn to help out.

But if these are minor inconveniences, it can get really heavy if you have to cope with the tensions within the other family, or worse still if you can't stand the way they handle their children. As when a precocious kid seems to get away with murder and you want to scream at it – only to get an indulgent smile from mum. Or you witness some heavy-handed punishment that makes you wince and immediately think of all those child-rearing manuals you devoured when it was your turn. All you can do is offer a weak smile and a glazed stare. Such moments do not a holiday make.

The honest point is that a holiday with friends is more of a commitment than simply booking into a hotel. So it is better to enter into such an arrangement with your eyes wide open. Anticipate as many potential disasters as you can, and then stop worrying and take them in your stride. If you're with real friends you'll sort it out. And there's nothing like a minor crisis or disagreement to get to know someone better.

JM

Afterthought

'What sweet little cakes,' exclaims the guest. 'Why, I've already eaten four!'

'Five,' says the hostess, 'but who's counting!'

LB

Holiday rows

Before she died my mother said regretfully, 'Here's some money for your holidays, boys. Take as many as you can because at my age you can't enjoy any.' So after she died we booked four packages and then cancelled the lot while I holidayed in hospital. Ma proposes but God disposes and He generates even more horsepower than her.

Not to tempt providence after I finished with hospital, we risked a last-minute package to a little *Schloss* (castle) in Hungary and wandered through woods speckled with wild orchids. I could get accustomed to *Schloss* life – and regretted my relatives had been Goldbergs not Hapsburgs.

There were problems of course not of our making. We had landed in cholesterol country, very delicious, very creamy, very porky – though cholesterol and Jewish tradition are probably my problem not yours. But one problem was of our making. We ended our holiday with a silly holiday row, like so many of you. Perhaps we saw too much of each other. Perhaps we were too tired. Perhaps I started thinking of the post piling up back home.

But also like so many of you, we made up pretty quickly because we'd been through that scenario before. Why be hoity-toity and cut off your nose to spite your face? Also

I read in the Gideon Bible beside my bed about Abraham's row with God over Sodom and Jesus chasing the money-changers out of the Temple, which made our little spat seem very small beer. And making up can be lovely.

Reading the newspapers in the plane back home I felt sick – not from pig and cream, but from the religious rows that were begetting new religious grudges which were begetting fresh hatreds that could corrupt and destroy the credibility of all religions. They were murdering the love they testified to.

I tried to puzzle out why religions can't learn to say sorry and make up like ordinary holidaymakers, but have to preach at each other instead.

One reason is that religious people have to be genuinely humble, not just ritually humble, to say sorry and make up, and many religions have become too grand for that – though God is as present in package holidays as in great cathedrals, and saying sorry and making up are what religion is about.

Another reason is megalomania. Religious people like me often forget 'we're only in sales not in management'. This is very true. You can often see pious people inflate, and hear their voices grow from diminuendo to crescendo and from normal conversational to denunciatory declamatory as they pass from being God's servants to being God. It's a special problem for ministers of religion who hypnotise themselves with their own sermons. This self-importance also comes from the robes we wear and the furniture we sit on. Robes, pulpits, rites, habits, are the theatre and can remove us from the still, small voice. I've noticed that as couples row with each other on holiday, they become theatrical too, denounciing each other like Sibyls and prophets and using the same rhetorical pulpit tricks.

These two chaps would have found it easier to make up if they hadn't become so grand. As little boys they loathed each other at school and only met decades later in a station on their way to an old boys' reunion. One had become a very grand admiral and the other a mighty bishop. The devil got into the bishop and he addressed the admiral. 'Stationmaster,' he said, 'where is the train for London?' 'In your condition, Madam,' replied the admiral, 'I would be inclined to take a taxi!'

Here are some psychological and spiritual exercises, which may help both holiday rows and religious ones.

- If you're spoiling for a row (lots of people know this in advance), warn everybody around for damage limitation.
- In prayer imagine yourself as a kitchen sieve and think of God as the water which washes your bad vibes through you and out of you, so that you don't just forgive but forget what you've forgiven.
- Try role reversal! You argue your opponent's case and let her or him argue yours. Very revealing!
- See God in each other!
- Measure the disproportion between the cause of your row and its effect. What are you really arguing about anyway?
- Toss a coin for who's right and let the other stand her or him a drink at the bar.
- Don't theologise your row – it's often hypocrisy and hypocrisy is what parts of the Talmud, and the Gospels, gun at most.
- Enjoy your row! Ham it up! (You should pardon the expression.)

- Perhaps rows will always be a permanent part of your relationship, so learn not to say things while you're making them which are unforgiveable. Learn to make up! Making up is very Godly – read your Bible where everyone is having rows with everyone else including the Almighty.
- If you can't make up and do say things to each other which are unforgiveable and can't break the habit, perhaps it's time to disengage. Consult a counsellor.
- Are you both tired, tense, worried or have an upset tum? Have a book with a happy ending to hand and a nice sandwich with a bar of chocolate.

LB

Some thoughts

Bear in mind that life is short, and that with every passing day you are nearer to the end of your life. So how can you waste your time on petty quarrels and family discords? Restrain your anger, hold your temper in check, and enjoy peace with everyone. (Nachman of Bratslav)

Rows wouldn't last long if the blame was all on one side. (La Rochefoucauld, *Maxims*)

Lord, help us not to despise or oppose what we do not understand. (William Penn, the Quaker)

Abraham gave hospitality to an old man in the desert. While they were eating Abraham discovered he was a idol worshipper and drove him from his tent in anger. In a dream that night God appeared to him and said, 'I have put up with that foolish old man for a lifetime, and you can't put up with him for a few hours!'

Family holiday

In theory it should be ideal. No work. The family together. A nice change of scene. Maybe sea and sun. 'Special activities for the children with uncle Mike and the Lolipop kids.' And sometimes it works out that way. But sometimes it doesn't. And the reasons are pretty obvious. Every family has its tensions and all the members have their own needs and wishes. At home we get by on routine and familiarity. Life is organised around the things we have in place, from food in the fridge to the telly in the corner to the place each of us can hide when we need to. But on holiday, in unfamiliar surroundings, in a couple of hotel rooms, the support system is missing. And the greater our expectations that this holiday will be 'fun!!!', 'pleasure' and 'rest' the tougher it is likely to be.

So a holiday with the kids is more of a challenge than we expect and it may be necessary to negotiate some ground rules from the start. So that everyone can get something of what they want for themselves, everyone needs to give up something for the others.

That sounds solemn but it's common sense. If mum wants to swim and sunbathe and dad prefers to sit in the shade with a drink and read and son wants to go inspecting the rock pools (or the discos if he's older) and daughter wants to pony trek (or date the waiters if she thinks she's old enough) things are set up for rows, sulks, tears, slammed doors, disappearances and mysterious illnesses.

To all of which there is no answer – only object lessons in the art of compromise. Making space for the needs of each (except for dating waiters when below a certain age) means setting aside a slot for dad's time and mum's time and kids' times while agreeing to share each other's interests together whenever that is possible as well.

Since it is often the mum who holds the household together it is only right and proper that dad take on the planning of his fair share of the days, including sorting out meals. And if what the kids want to do is excruciatingly boring or patently absurd, that's presumably how they view most of what they see adults get up to, so it's only fair to give them their revenge.

So when we're sitting in an amusement arcade, in front of rows of noisy computer games, deafened by heavy metal noises that pass for the music of a younger generation, we can bask in the glow of self-righteousness at yet another minor sacrifice for the sake of the family as a whole. And think what a delightful break it will be to be back in the fascinating routine of the office or the kitchen next week.

Family holidays are a chance to see how the family develops year after year, though it may be hard to see this, or appreciate it, at the time. But decades later snaps or home movies are marvellous reminders of how it once was – when the kids were young and life was simpler. That is when we forget the fights and difficulties we had at the time – or realise just how silly they were, and how they were part of what family relationships are all about.

For kids have to test themselves against their parents as they establish who they are, what are the boundaries they have to discover and push against? Sadly it is often only much later that we understand what those struggles were really about.

Some advice

Since the time we are together before the children go their own way is so short, when the family holiday really is a family one, it does help to think the holiday through ahead of time. That way we can make sure that there are activities all of us can share, but also just enough private space so that we can stand back from time to time, to enjoy the rough and tumble and try to understand it. Family life is all about the extraordinary privilege and responsibility of nourishing the hearts and minds of another human generation.

So don't wait for the snaps and home movies to feel nostalgic; don't wait till it is too late to get caught up in the mini-dramas, the extraordinary fun and the constant surprises of a grown-up kind of loving. Cherish the moment!

JM

Past wisdom

> Never promise a child what you do not intend to give. (Talmud Succah)

> Do not threaten a child. Either punish or forgive him. (Talmud Semachot)

Parents should never tell a child they will give him or her something and not keep their promise. Because in that way they teach the child to tell lies. (Talmud Sukkah)

A father complained to the Baal Shem Tov, the spiritual teacher, that his son had forsaken God. 'Rabbi, what shall I do?'

Replied the Baal Shem Tov: 'Love him more!'

There is one thing in prayer that we all need so greatly and which children can already learn: a few minutes each day of peace and quiet with oneself, at least a minute fragment of the day to listen within oneself, attending to the voice of God. (Ellen Littmann)

Never enter your house with abrupt and startling step, and do not behave so that those who live under your roof dread your presence. (Eliezer ben Isaac)

Present experience

These reflections and anecdotes on holidays with children come from my friends Derek and Gabrielle Stanley, who have children – I don't have any.

LB

1. First get your terms right. Holiday does not mean 'rest' or 'relaxation' for the parent. That does not occur until children are old enough to mix a gin and tonic, by

which time they don't want anything to do with parents and go off on their own!

2. In the PMT era (Pre Mobile Telephone), travelling in convoy with friends/relatives, when on the motorway, how do you let the car behind know the child needs to stop to use the potty? Well, don't brandish said potty through the sun roof! Wind-speed factor will whip the potty out of the firmest grasp and send it hurtling towards the windscreen of the following vehicle. Only do this if you want to fall out with the rest of the party on the first day of the 'holiday'.
3. Going down to the beach means three hours hard digging in the sand for the adults.
4. Don't take an interesting book.
5. If the beach has a lot of pebbles don't be surprised when you look up and see half the beach is now contained in your son's swimming trunks. It is not a form of bowel disorder!
6. Monument bashing – always a crashing bore for children. However, sometimes they can draw the adults' attention to something interesting. Our bored 12-year-old son looked for somewhere to sit down whilst we were trailing round the Palazzo Pitti in Florence. Yet another room full of Madonnas and pomegranates! Bored son found seat temporarily vacated by security staff and gazed upwards. Magnificent painted ceiling depicting some gorey battle scene, with diaphanous half-clad maidens floating around. 'Quite nice ceilings here,' quoth bored boy to industrious adults who up to that point had not lifted eyes much above head height. We never looked down after that!
7. If the children misbehave then punish them by trying

out schoolboy French on startled waitresses. Even worse, try making a joke in the foreign tongue.

8. Do not be surprised when children come out in mysterious unidentifiable spots for no good reason at all. This usually happens on a Sunday when everywhere is closed and all doctors are on holiday as well!
9. After sleeping all day, bored teenager sets out on bike ride when the temperature is still about 30°C. Returns an hour later, dusty, speechless with exhaustion and heads for swimming pool, pausing only to remove walkman and earphones before plunging in fully clothed and shod!
10. Try seeing in the New Year by persuading everyone to go to bed for a nap in the early evening with a view to getting up just before midnight. Forget to set alarm clocks. This way everyone feels great at 9 a.m. on New Year's Day!
11. Do your best to persuade someone else to take them on holiday!

Working holiday – dancing the Gay Gordons

I hadn't danced for years. Not since the sixties. I remember exactly when I stopped. It was in the sixties. In the middle of a dance on holiday, my partner suddenly unhooked herself from me and started doing her own thing. Feeling a fool trying to dance on my own, I retired to my seat in a huff and never took to the floor again, until, that is, when I whooped, yelled and pranced in the Gay Gordons.

This came about on another holiday, a working one. Like many professionals I sometimes try to combine a holiday with a conference. I give a lecture and they provide my fare and stay. Some of you will exclaim at my luck but combining them isn't easy. Sometimes there's not much holiday to it and you're conducting a problem clinic far into the night.

The friendliest one I ever went to (and the most pious) took place in Scotland. That's when I dared to dance the Gay Gordons because I was so relaxed. The university where it took place was reassuring. The halls of residence were full of fellow oldies like me, filling the holidays of their sunset years with folk dancing, origami and Mary

Queen of Scots. Americans with Jacobean pretensions pored over tartans, and pleasant Japanese photographed me, each other, each other and me, and each other and every sporan in sight.

At breakfast we compared the conferences we were attending. It was rather like gabbling our different confessions of faith or the conventions we used at bridge. A transatlantic lady who, like me, was into oatcakes slathered with Dundee marmalade, asked, 'Are you researching clan tartans?' 'No,' I replied through a mouthful of my marmalade, 'I'm at a conference for lesbians and gays.' She stopped masticating, mid-mouthful. 'Well, what do you know!' she muttered.

To reassure her we were no lewd lot, I added, 'They're all Catholics, very pious and prayerful.' 'What, real Romans?' she exclaimed. 'Oh, yes,' I answered proudly 'cradle Catholics too, not just converts! Why, we've even got two priests and a nun.' She digested this information along with another oatcake. 'Are you a friar or something?' she said, probably because I looked too dishevelled for solemn monk status. 'A something.' I said. 'A Reformed Jewish rabbi.' She clapped her hands and exclaimed delightedly that's what she loved about holidays. You never knew the surprises you would experience on them or who you would meet. Well, now she had heard and seen it all.

But she hadn't really! She should have seen our conference at the ceilidh, with our lads in their kilts, frills and dirks and our lasses in sashes, skirts and trews, kicking up their heels in the Gay Gordons. It was the first time I'd danced for years as I've said and I'd rarely felt so at home and happy. Two clansmen assured me I could wear a kilt too – Blues really existed as a subclan on some remote

Highland island. They also said I had the right rump. O those Scots! A little economical with the truth maybe but how courteous *quand même*!

My breakfast friend interrupted these reminiscences by suddenly asking me, 'So what do you get out of it, rabbi, seeing you don't even share the same scriptures?'

She was right. But all gays have a problem combining the truths of tradition with the truths of their own experience. Some gays are bitter and short-circuit the problem saying, 'Since my faith has no place for me, I've no place for it!' and never go to church or synagogue again. But the people I met were following the harder path by piecing together the public scriptures from the past and the private scriptures of their own lives.

The latter are sort of scriptures, too, because God didn't stop speaking at some date in the past but continues His work in us, refining His revelation. That's how slavery was not merely ameliorated but abolished and women have not only acquired status but are now moving on towards equality. God's continuing revelation has stopped us torturing old women as witches, permitted us interest on our building society accounts, taught us to respect other faiths not fight them, and it's why majorities no longer burn minorities, theological and sexual, but tolerate them, even if they don't yet accept them.

I was pleased I had worked it out and turned to enlighten her. But I had been pondering in silence too long and she had gone to hit today's Whisky Trail. But the Japanese were still there. One of them asked me politely why Englishmen like me oppressed the Scots, denying them their freedom. Through the window I pointed to some brawny, kilted chaps bounding up Arthur's Seat. And I replied equally politely that any such subjugation would be most imprudent from a puny, pusillanimous southern Britisher like me, of the Mosaic persuasion.

While he looked up these exciting new words in his upside-down dictionary, I blessed God for freeing me from my ghetto. I had begun to think of the world as one big Jewish problem, and in the Scottish fresh air realised again a truth I'd almost forgotten, that more basic even than being Jewish or English was being a human being. My transatlantic friend was right. 'You never knew the surprises you would experience on holiday.'

She was so right. I felt exhilarated by my fresh freedom. And that's how I came to whoop and yell in the Gay Gordons that night.

LB

Questions

Do holidays make you feel free too? Or are you too stuck into your own past? What have you learnt from them? What would you like to learn from this one now?

Some thoughts on dance

> Rabbi Nahman of Bratslav tells of a child watching his grandfather dancing and asking him why he danced all the time. The grandfather replied, 'You see, my child, people are like spinning tops. They can only keep their dignity while they are in movement.'

> Rabbi Berechiah and Rabbi Helbo said: In the world to come God will lead the dance with the righteous.

Working holiday – dancing the Gay Gordons

> The righteous will be on one side and the other, with God between them; and they will dance before God with vigour and point with the finger and say: 'This is our God forever and ever, who will lead us beyond death.' (Song of Songs Rabba 1:3)

> Let them praise God's name through dancing,
> making music with timbrel and lyre!
> (Psalm 149:3)

In the privacy of your room dance some steps yourself. It will free your body and your mood. Shimmy!

LB

Holiday reading: Going to bed with Gideon

Holiday reading

One of the joys of a holiday for me is the chance to leave behind the usual solemn books I have to study and lose myself in another kind of writing. The racier the style the better. A café on the sea-front or a park bench, a paperback novel short enough to be devoured in a day, and I am in readers' heaven.

It started as a kid, and part of the attraction was probably the simple mania for collecting. To own a shelf-ful of those colourful paperbacks about some hero, and the thrill of adding one more to my collection, was probably as exciting as reading the books themselves.

I worked through Agatha Christie at an early stage. I'm still fond of Hercule Poirot, but I got bored with those endless house parties, the inevitable setting for the classical English detective story.

Edgar Wallace was less genteel and the Saint fed my fantasies. Then I discovered another kind of writing and my fate was sealed. Raymond Chandler came into my life as he said the thriller should happen – when in doubt, have someone come bursting through the door shooting a gun. The fast pace, the smart one-liners, the feeling that this was reality and not just a delicate case of poisoning in the vicarage, totally gripped me. From then on I was hooked on all the wise-cracking, hard-boiled, private eyes, cynical but with a deep-down integrity. They operated out of a dangerous but glamorous place called LA. They were streetwise, romantic but unencumbered and fell in love with their female clients only to be betrayed by them on the last page but one.

So what is it about these books that is so attractive? Part of it, of course, is solving the mystery itself. It is fun trying to figure out what is going on, especially as we peep over the shoulder of the detective who very kindly sees for us, interprets for us, gets beaten up for us and fights the last battle for us. *He* may be guessing much of the time what's going on, but at least *we* know we are heading towards some kind of resolution by the final page.

But there is another great security that underlies all these series. We know that in the end the hero will win out and live to fight another day, however dangerous the dark streets he walks and however hair-raising the particular threat posed to him by the baddies. After all, even Sir Arthur Conan Doyle was forced to resurrect Sherlock Holmes because of public demand after he had successfully killed him off. We want our invulnerable, unconquerable heroes and the reassurance that they will come back fighting in the next book.

Perhaps that is at the bottom of it all. Real life is too full of uncertainties. We are not very good at picking up the clues to our own story and are far more likely to go chasing after red herrings. Nothing we own is permanent; our relationships are complicated; chance can turn our lives upside down in a moment, and on top of it all at the end of the day, like our hero, we have to pay our dues, though in our case it is not usually a debt of honour but our income tax. But our detective heroes, even if they struggle to make ends meet, even if their love life is a disaster, even if their world seems to fall apart about them, and even if they are at odds with the cops and the gangsters at the same time, we know they will win through. So for a few moments we can share their disasters and triumphs in the certain knowledge of a happy end.

True, our own life may not be as well written as a detective story. Most of the time we can only know after the event what was really going on, and not always then. So our spiritual task is to learn to live with the mess. There are no quick-fix solutions and there's no guarantee of a story-book happy ending. We cannot write the plot in advance but we can decide how to live with whatever comes our way. We have to ride the punches and struggle back on the trail. So maybe we can be the private eye in our own personal novel and try, like the best of them, at least to keep our integrity whatever fate may throw at us. After all, we're around for quite a long time – and till it's all over there's always room for one more pulp novel about our adventures on the great bookshelf of life.

JM

Anthology

Days are scrolls, write on them what you want to have remembered. (Bachya Ibn Pakuda)

A book is a marvellous companion . . . It is inanimate, yet it talks . . . You will find no more faithful or attentive a friend in the world, nor any teacher more effective . . . It will join you when you are alone, accompany you in exile, provide light for you in the dark, and entertain you when you are lonely. It will do you good and ask for nothing in return. It gives and does not take. (Moses Ibn Ezra)

JM

Holiday friends

These books have become my holiday friends.

On Not Being Able to Paint by Joanna Field, because it showed me I was able to paint, though I couldn't get a likeness and couldn't draw for toffee.

Cranford by Mrs Gaskell, because it's the kindest novel in our language.

Other Men's Flowers by Field Marshall General Wavell, because it contains my favourite poems.

A Book of English Belief by Joanna Mary Hughes, because she embodied the best of England and the writers she quotes became her spiritual friends.

The regency novels of Georgette Heyer, because they all have happy endings and there are not enough of them around.

Perhaps I ought to take Proust, *The Cloud of Unknowing* and Ivy Compton Burnett instead but the above are what I really want.

LB

The Bible by your bed

At the end of the nineteenth century some American businessmen, who wanted to fight the good fight against irreligion and vice and who knew the temptations of travellers, arranged to place a Bible in hotel rooms and boarding houses. You will probably see one in the cupboard by your bed. You may have thought it was a telephone directory. If you want to know who Gideon was, satisfy your curiosity by looking up chapter seven in the Book of Judges.

If you have run out of romances and detectives and feel alone, abandoned by friends, spouses and lovers (a common feeling if you wake up in the middle of the night), then why not turn to that Gideon Bible to see if the journeys in it interest you or, more important, are relevant to your problem and condition.

Bible reading and Bible searching are not usual now, though they were in your grandparents' time and in countless generations before them. There are other good reasons for dipping into that Bible. First, it is full of characters, many of them more passionate and less pious than you and certainly more complex than those in magazine romance. You could learn from their life experience. Another reason is that some of it is very beautiful and well suited to a bedroom. Some of the Psalms (the first part of Psalm 23 or Psalm 131, for example) might lead you into a peaceful sleep without tablets. Yet another reason, some parts deal with your conscience and soul which otherwise don't get much of a look in on holiday, but are necessary for a holiday lest it leave a bad taste or bad memory behind.

There are difficulties of course. There are the bits of formal archaic language. Don't worry about them! You are not mugging up your Bible for a Sunday-school test, but browsing through it to enjoy it.

Do you have to believe it all? That's up to you. Some say 'yes' and some say 'no'. For me personally it is a mixture of what actually happened, what people would have liked to have happened and what they thought was the significance of what happened. You can sort all that out later.

You also have to delve into the text and chew it over in your mind. The books of the Bible (it is really a library not a book), like most ancient books, weren't intended for fast reading. Anyway, here are some travel stories from the old Hebrew scriptures unpacked for you. After reading them you might want to go on and unpack your own.

They are about journeys not holidays because modern secular holidays hadn't been invented yet, though holy days and rest days like the Sabbath, and festivals, were well known and generally observed.

I suggest you read the extracts in the following chapters in your Gideon Bible first then read the unpacking and then back to the Bible again. The Bible can get very addictive.

LB

An oldie starts a new life (Genesis 12)

A friend of mine turned seventy and decided it was time to give up holidays and travelling. After all she was now on her way out! When you get into that kind of mood try reading about Abraham whose life only really started when he was seventy. I can't say he had a fun time, but it was never dull and he never stopped searching.

Abraham is not the most obvious person to choose to start a revolution in the human heart. When we first meet him in the Bible he is already seventy-five years old. He seems to have grown up in the shadow of his father Terach since he still travels with him when Terach decides to leave their home in Ur of the Chaldees to go to the land of Canaan. They make a stop-over on the way in Haran and get stuck there. Abraham has to leave his father behind to finish the journey.

Abraham cannot have been a very happy man. His wife Sarah was barren, and that was considered a particular tragedy in biblical times. No children meant no one to look after you in your old age or to keep your memory alive after your death. So when God called Abraham to complete the journey to the land of Canaan, he really had nothing to lose. Old though he was, this might have seemed like a last chance to make a new beginning even now in the autumn of his life.

It could have ended up in total failure. Or Abraham could have seemed just as silly or pathetic as Don Quixote – another old man with a strange vision, setting off on a romantic journey to rescue damsels in distress and tilt at windmills.

Instead Abraham fathered a people that is still around today, Judaism sees him as its founder and he is regarded as a key figure in Christianity and Islam.

The call he gets from God is not really a command, though most English translations make it seem so: 'Go!' or 'Go now!' But the original Hebrew has two words which make it more of a request: 'Go – for your own sake.' He is to leave his old world behind and take a chance on a new one.

So Abraham went and arrived in the promised land. But here reality intrudes. The times are hard, there is famine in the land. And just when he is in trouble it seems that God is nowhere to be found. The voice that told Abraham to go is now silent. So Abraham gave up on God for the time being and used his own initiative. He journeyed down to the land of Egypt in search of food.

Abraham must have felt betrayed. After all he had given up everything for this journey of a lifetime. Perhaps he thought there was a 'pot of gold' waiting for him at journey's end or at least a 'happy end'. Instead he found himself stuck in yet another foreign country, living off his wits. Even so his heart was still set on that 'promised land' and he headed back at the first opportunity to try it again.

His departure from Egypt was without much dignity

since he was escorted out under armed guard! He'd made a detour, but he'd also passed some kind of test by being willing to trust God and try again.

There is another way to understand his call. ' "Go" where I send you,' says God, 'but the rest is "up to you". You will have what you need for your journey, but you will have to find your resources within yourself. And though we will meet now and then on the way, it will be at the times of My choosing,' says God, 'not always when you want it.'

All of which seems to belong to a holy day more than a holiday. But anyone who's had a flight cancelled or had to change a booking because of some last-minute crisis knows that we are not in charge of the universe. 'Man proposes, God disposes.' But whether 'holy' or only 'holi', we are still in charge of how we deal with the things that come our way. So when things don't work out as they should on holiday, remember ancient Abraham, without benefit of bus pass or travel insurance, packing up and heading off knowing that somewhere unexpected along the way God was waiting to meet him.

JM

Jonah and the big fish (The Book of Jonah)

If you like tall stories, few come taller than what happens to the hero of the Book of Jonah. And if your hotel accommodation is a bit cramped, compared to three days inside a fish you're living in a palace!

It is probably the most famous journey in the Bible – Jonah's ocean voyage which ended up with three days spent in the belly of a big fish. The Bible doesn't actually say what kind of fish it was but the story is well known as 'Jonah and the Whale'.

God told Jonah to go to Nineveh, the most evil city in the world at that time, and tell them that because of their wickedness God was going to destroy them. The Bible does not actually explain why Jonah did not want to go, so it is left to us to try to figure it out. Or else simply imagine to ourselves the last place on earth we would ever wish to visit. Then we begin to understand him. Whatever Jonah's reasons, he headed out to sea in the exact opposite direction from Nineveh, in fact to the other end of the known world, to Tarshish, not far from Gibraltar or modern Torremolinos. If you are running from God you cannot go far enough away.

The rest of the story is also well known. God sends a storm onto the sea, the sailors find out that Jonah is the cause and ask him what to do. He could have said, take me back to shore and I'll go to Nineveh. Instead he asked them to throw him overboard. But Jonah did not have to involve the sailors in his final choice, he could simply have jumped over the side by himself. Maybe he did not want to make that final decision himself and left it up to them. The sailors clearly did all they could to save him and themselves, but in the end had no choice in the matter, and overboard he went.

There waiting to meet Jonah, was the great fish where he stayed for three days and three nights.

The early commentators were very interested in what happened during those three days since the Bible says nothing about it. Some thought that the fish took Jonah on a conducted tour of the sea. Others wondered why it took him three days before he finally prayed to God, and they weren't very impressed with his prayer when he finally spoke. It did not say anything about how he got into the fish in the first place and there was not a word of apology to God. Certainly the fish did not think much of it and threw up – which is how Jonah came to be saved in a very undignified way.

Jonah does go to Nineveh, walks part way into the city, says five words and walks out again. Then he sits outside waiting in the hope that the city will be destroyed even though he knows it won't be. Jonah comes across as rather childish.

In the end of the book God asks him if he could not feel just a little bit of pity for Nineveh, but Jonah doesn't even answer. He remains stubborn to the end.

Despite all that, I'm fond of Jonah because I can see so

much of myself in him. If I have to do something that I do not really want to do, even though I know it is right and I should do it, I'm always tempted to run as far away as possible. Luckily there are usually a few friendly 'sailors' to set me back on the right track, though I'm still waiting to meet my own big fish.

Further thought

There was once a severe flood warning and the army turned out to warn people. But one pious man refused to leave his house. 'God will save me!' he said with all the certainty of faith. Well the flood came and the streets were filled with water. A boat came by to rescue the man who by now was standing on the first floor balcony. He smiled knowingly at them and called out: 'There is no need to worry, God will save me!' The waters rose higher and finally the man was forced to climb onto the roof. A helicopter passed and lowered a rope to haul him to safety. Again he called out, 'I have no fear, God will save me!' But the waters rose even higher and he was drowned.

When he got to heaven he was quite upset and demanded to know why God had not rescued him. But God was quite puzzled. 'I don't understand. I thought I sent the army, a boat and a helicopter.'

JM

Travelling on a donkey (Numbers 22–24)

Sometimes things just go wrong on a holiday and we feel the need to find some kind of explanation just to make sense of it all. That's the time to read about a man called Balaam and his donkey.

There is a phrase that pops into my mind at odd times just when I don't want it. And like all such phrases, the more I try to suppress it the more it sticks in my mind. I think it was coined by the Scottish writer Robert Louis Stevenson, and it says: 'It is better to travel hopefully than to arrive.'

It is a marvellous phrase, and I'm sure it is true. Since Stevenson wrote a book about travelling with a donkey it is probably backed up by his own personal experience. The trouble is that on at least a couple of occasions when it has occurred to me during a journey, something has gone wrong shortly afterwards. So I end up being very superstitious about the phrase whenever I travel, just in case it occurs to me. Of course it is silly to take it seriously, and there have been ever so many occasions when I did think of it and nothing went wrong. But those are the occasions we tend to forget, and after all, you never know . . .

Thinking about this phrase, and especially Stevenson's donkey, reminded me of another special donkey from the Bible – the faithful donkey on which the prophet Balaam rode.

The children of Israel have just crossed the Sea of Reeds and the fame of their God has spread before them. The surrounding nations are scared, and as the children of Israel head towards the promised land, King Balak sends for the sorcerer and prophet called Balaam to come and curse them. As the king himself points out, whoever Balaam curses is well and truly cursed.

Balaam goes to ask God for permission but is told in no uncertain terms by God that he may definitely not make such a curse. But Balaam simply does not want to hear this and goes off with Balak's delegation to earn a large fee and enhance his reputation. At which point an angel and Balaam's talking donkey take a hand.

The donkey sees an angel with an upraised sword standing in the way ready to slay its master. Balaam is so blinded by his wish to curse the children of Israel that he cannot see the danger ahead of him. The donkey stops to protect its master from the angel, but Balaam beats it to make it move on. The same pattern is repeated, and the next time Balaam bangs his knee on a wall as well and hits the donkey even harder. Finally the donkey has had enough and opens its mouth. It points out that it has served Balaam faithfully for a very long time and has never let him down before, so Balaam should trust its judgment and stop hitting it! At which point Balaam's eyes are finally opened and he sees the angel and realises that there is something very wrong in what he is doing. He cannot curse the children of Israel as he would have wished. In fact he ends up blessing them instead, much to the fury of King Balak.

Balaam travelled hopefully but almost never arrived. His story seems to be about an outer journey but it is really about an inner one. Balaam was torn by conflicting wishes within himself. His reputation depended on his doing a good job of cursing, and he may have had some kind of personal animosity against the children of Israel as well. But he also knew that this was one job he should not have taken on since God was against it.

Now, it is not every day that we encounter angels and talking donkeys or for that matter receive messages from God telling us to do the opposite of what we desperately want to do.

But Balaam's story gives a helpful warning. When our expected journey is constantly being interrupted; or if we find ourselves hurting the people we love or trust, it is time to stop and take a look at what is going on in our life. Perhaps we have taken a wrong direction. Perhaps we are no longer being true to ourselves and cannot see what is staring us in the face. And because we cannot sort out our inner problems we are taking them out on someone else. So instead of going round cursing others we should take a look at ourselves. Maybe the people we are cursing are really on our side and deserve a blessing instead.

If we do travel hopefully, then wherever we arrive may just turn out to be the place where we ought to be.

JM

The journey of a young woman in a man's world (The Book of Ruth)

Being alone means you have to do that bit more to find company and make a place for yourself. If it gets you down try reading the Book of Ruth, the story of a woman who managed to make something out of a lonely and difficult situation.

This century has been described as the century of the refugee. There are more displaced people in the world because of famine and war, natural disasters and human-made tragedies, than at any time in the past. People are on journeys, not because they wish to, but out of necessity. They have to abandon homes and families and security and familiar surroundings to try to find a place of refuge somewhere else. And that somewhere else is often indifferent to their needs, at worst unwelcoming and hostile.

Though the scale of such displacement is new, the problem is as old as humanity. The Bible tells of the relocation of whole populations. Many were forced to move because of war or famine. But some left of their own free will, going from a place of slavery to find a new home, as when the Israelites left the land of Egypt. The Bible also tells us some individual stories about those who undertook a journey to a strange land. One such case is a woman called Ruth from the land of Moab, whose journey turned out to be so important that an entire book of the Bible was named after her.

We meet her through her family connections. She married into an Israelite family that had been living in Moab. They too had been driven away from their home by famine. Then tragedy struck and her father-in-law died, as did her husband and brother-in-law.

Ruth was faced with the choice of returning home to her family in Moab or staying with her mother-in-law, Naomi. It must have been a very difficult choice. Naomi was depressed, having lost her husband and both her sons. All she wanted to do was return to her native land, alone, unencumbered by anyone. When her daughters-in-law offered to return with her, Naomi told them to go away and leave her alone. One left, but Ruth decided to stay.

The story that follows looks like a fairy-tale ending. Ruth returned to the land of Israel with Naomi and went to work in the fields of a relative called Boaz. He soon recognised her qualities. At Naomi's urging Ruth took the risk of visiting him at night at the end of the harvest where he slept on the threshing floor. She was wearing a wedding dress, but Boaz could have simply taken her on the spot and abandoned her. The Bible doesn't say what happened that night and leaves it to our imagination. But theirs wasn't just a holiday affair. Ruth and Boaz married and Ruth turned out to be the great-grandmother of King David.

The journey of a young woman in a man's world

For Ruth this was no fairy tale. Like other refugees she had to put up with suspicion, fear, prejudice and downright hatred. So reading her book is a way of reminding us that a holiday journey is the kind of luxury that few can afford in our troubled world. That may help us appreciate a bit more our own good fortune, so that we do just that bit more when we get home for the refugees that are not as lucky as Ruth. Like Ruth herself, we can help change the plight of a defenceless refugee into a symbol of hope.

JM

Re-entry

Memories

Do you remember that place you once came across as a kid? You were on holiday with your parents, so most of the time it was boring. Still, you got an ice-cream almost every time you played up and your first real pocket money just for sweets. But the rest really was boring, following them around to boring sights and boring shops. Then one day you went on a trip a bit off the beaten track. Up a hill that felt like a mountain, but you kept up to show how tough you were and to spite your younger sibling who complained even more than you and at one point had to be carried because of a sprained ankle. (Some sprain! Didn't stop that one from dancing all evening. Clearly just an attention-getter.)

And then you turned a corner and found it. An enormous castle soaring up to the sky with narrow slit windows for shooting arrows, or maybe the special trap for pouring burning oil on invaders. Or was it instead a waterfall that danced down the mountain, mist rising from the turbulent pool below? Or a well they told you had magical powers if you tossed in a coin and made a wish, and you all had a go and made a wish together and maybe even some of them came true. Or that field of heather stretching into infinity, ringed by distant blue hills, and how strange the few cars looked scattered across this charged landscape.

Whatever it was, it transformed that holiday into something extraordinary. You have only to close your eyes and think about it and it leaps again into view as clear as ever before, clearer because you had not yet learnt that protective detachment, some say cynicism, that censors out instant reactions, too much pleasure, in case . . .

So one day you decide to go back there, to try to find it again. Perhaps to share it with your own kids or someone special in your life now. Perhaps they too can see what you saw, feel what you felt. You want them to share it, or maybe come closer to you as well. It becomes in your mind a kind of pilgrimage, though you know only too well the risks. It cannot be the same because you are not the same. It will be smaller now because you are larger. Other experiences, other castles or meadows or wells have jaded your palate. Maybe it isn't there any more – the developers have moved in and put up dwellings so that a privileged few can enjoy the view, or to make it more accessible, less rare. It might even be, as a kind of joke waiting to mock you, that you cannot find it. It wasn't on *that* holiday, in *that* place, that you saw it, and you were only a kid at the time so your memory is sure to be faulty and it's too late to ask your parents as they're dead and your sibling has no recollection of it at all. Maybe it never meant anything to your parents, so with the best will in the world they cannot respond to your insistent questions. You almost begin to doubt it ever happened at all – but only for a moment. Some things cannot be lost – and maybe next time you go looking you will have more luck. You'll take that other turning, plan the route better.

Memories

In your heart of hearts you know it is too late. Not that you couldn't return, but what is the point? The memory has survived either the real place or the inability to find it. Something broke through, a sense of wonder, awe. It touched the artist in you, or the soul. We use different words that make sense in our time and for our culture. But however we express it, that moment was truly there and so it can be searched for again and again. That is the essence of hope. And maybe, just maybe, one such moment is enough for an entire lifetime.

Our religions and all spiritual traditions are full of such wondrous places – where miracles happen or some light is revealed and we have the chance to peek around the screens behind which God hides. Some places can be located and become the site of pilgrimage and piety, but also property values and, all-too-often, holy wars. Other Utopias stay for ever out of our reach – somewhere at the end of the rainbow or in a magical kingdom at the bottom of the sea. Try to think of your ideal place and what it would be like. What would it take to build it, at least some part of it, in the world around you? What we can imagine will always nag us with the hope of something better yet to come. And that too will have to sustain us during the long, long wait that is our life.

JM

Returning

Returning isn't an easy time to manage. It's a time when relationships are tested, when a couple need to help each other, not be sharp with each other. It's never easy adjusting to another sort of life. Yes, you did precisely that when you arrived! But then you had time to work things out. And there's no time when you come back. Letters will be awaiting you and bills and urgent messages on your answerphone. There will be no rep to complain to either. And there'll also be the jet lag if you're coming from faraway places.

Sometimes airport delay makes it worse. If you've cut it rather fine, what about your job? I chatted to two workers from the Midlands already delayed for eight hours who were drinking their way out of that anxiety.

Because returning is an in-between time, that's when you mislay things. Not lose them, mislay – real losses and thefts are not that common. Take it gently, and before you panic, check your pockets systematically and carefully. Casual holiday clothes have pockets in unexpected places.

Remember the nice things to hold on to – what's going on in your garden, distributing gifts to neighbours and anticipating the pleasure they will give, and knowing that holidays will come again.

These tips have helped me. But in the hurly-burly of holidays they might seem like counsels of perfection. When you book your holidays try and arrange a clear day before you go back to work. Then you can potter and make friends with your house all over again, just as you did with your hotel room when you arrived. You need time to open and shut drawers and make cups of tea so that you can calm down while taking your return on board.

Also think ahead. Try and arrange your return from the airport before you go. If you haven't got a car, is there a convenient bus or train, or best of all is there a neighbour, friend or relation to pick you up? That's when you know who your friends are. But don't spring it on them at the last minute. Also make sure you've kept some UK change or have a phone card ready. Don't use all your change on the way out.

I once suggested to some nuns who were asking themselves what new work of service they could do in the new age, now that they were no longer running schools and hospitals, that an order which devoted itself to the needs of tourists, and travellers is needed more now than it ever was in the Middle Ages.

I've seen mothers with crying babies who desperately needed a hand in the departure lounge while they waited and waited and waited for their flight to be called. And there were couples who were near breakdown of relations because they both needed help against the tetchiness and bad temper that were covering their inner anxiety and bewilderment. Some friendly 'there-there' assurance would never come amiss or be resented.

So far no nuns I know have taken up my suggestion.

They are working for justice issues. But we are all God's hands in the world and the work we are first called to is what lies in front of us not thousands of miles away.

But back to the departure lounge when you return. Take normal precautions. Pack in your hand luggage a book with a happy ending. This is the time for comfort reading not culture. Remember you have to love yourself as well as your neighbour. Indeed, as the Bible says, you can't do the latter without the former. Have some jokes at the ready to cheer up those around you. If they don't work, don't worry. It may not be your jokes but them. At least you've tried.

Don't drink too much on your way home, you don't want a headache with your heartache.

If you're one part of a couple, be sensitive to your partner's needs. Think about what they are. Now isn't the time to say whatever comes into your mind. Steer clear of criticism, however justified it may seem to you. Award yourself a Brownie point for every tiresome truth you didn't say or suppressed. Be especially kind about credit card spending, suspicions of infidelity even if only mental, tripping you up in line dancing, and lots of niggling things that can hurt. Now is not the time to unload. Be a pal!

In these situations God comes in useful. You can unload on Him instead. How do you start? Well don't compose prayers. Start off with anything you remember from Sunday school – the Lord's Prayer, the twenty-third Psalm, 'the Lord be in my head . . .', anything which brings the divine into the picture. And if you're so annoyed you can't be nice for your partner's sake, then do it for God's sake. If you concentrate on Him, then the pluses and minuses of life fall into place.

Lots of prayers aren't answered, I know, and so does

everybody else. But in my experience one always seems to work. It's when I've prayed for that bit of extra strength to do what I have to do. It's stood me in good stead when I've had to cope with blood, when my partner's walked out on me or vice versa, when I've lost my credit cards, when I'm in an anxiety state. Counting my blessings also helps, which are more numerous than I thought.

It's a help if you rehearse in advance all the things that can come upon you – the delays, the queues – before you go. Then they don't come upon you unexpected and you've got them in some kind of proportion. This is a good tip to round off your holiday and also a good tip when you or someone you love rounds off your or their earthly life, as I found out in hospital when my mother came to visit me.

She looked terribly sad, which seemed only fitting, and then burst out, 'O Lionel, what shall I do? I'll have to say all those Hebrew prayers and they'll blame me for all the mistakes. And I'll get everything wrong.' Befuddled by tablets at first, I couldn't make sense of what she was saying. Then the penny dropped. She was rehearsing in advance, just as I'd advised, all the problems attendant on my demise – and me, her only son too! I giggled. 'Don't worry, dear,' I said weakly, 'I'll make sure there's a nice rabbi who'll sort everything out for you.' My mother then leaned across me and kissed me tearfully and passionately. 'O Lionel,' she said, 'you are lovely!'

LB

A suggestion

If you feel sad that your holiday has ended, there's a lot of uplift in little things, which the French call *petits soins*. Like making soup, polishing shoes and matching single socks. The next season's holiday brochure also helps.

Here's something sweet and something comforting and something homely (I've never had it abroad). Toast slices of bread. Don't burn them! Melt together in a saucepan four tablespoons of butter, or, better for you, low cholesterol margarine, four or five tablespoons of brown sugar and a heaped teaspoon of powdered cinnamon. Spread the syrup over one side of the toast and grill till the syrup bubbles merrily. Remember hot sugar is hot, so handle with care! To be relished with feet up, a strong cup of tea and your radio.

A simple blessing

Blessed are You my Creator who has looked after me, and watched over me and brought me back home.

Now close thine eyes and rest secure,
thy soul is safe enough, thy body sure.
He that loves thee, He that keeps
and guards thee, never slumbers never sleeps.
The smiling conscience in a sleeping breast
has only peace, has only rest.
The music and the mirth of kings
are all but very discord when she sings.
Then close thine eyes and rest secure,
no sleep so sweet as thine, no rest so sure.

These lines were written by Francis Quarles who lived in the first half of the seventeenth century just before Britain exploded into civil war. The inner peace of his poetry works even if your conscience isn't so clear. I vouch for it!

LB